STREET SUPERCHARGING

BY PAT GANAHL

STREET SUPERCHARGING

S-A DESIGN BOOKS
515 W. LAMBERT, UNIT E
BREA, CA 92621-3991

EDITED BY LARRY SCHREIB
GRAPHIC PRODUCTION BY LARRY ATHERTON

TECHNICAL DRAWINGS BY JAMES DENNEWILL

ISBN 0-931472-17-2

CONTENTS

CHAPTER 1
THE BASICS OF SUPERCHARGING

A 16-plug, dual distributor, 426-inch Chrysler hemi is a plenty impressive piece of machinery, but a big, whining, polished 6-71 supercharger adds a whole new dimension to such a motor—in performance as well as looks.

SUPERCHARGERS—EXOTIC OR PRACTICAL?

No single performance modification to an internal combustion engine is more practical or effective than the addition of a supercharger.

The foregoing statement may sound like typical technological bragging or advertising hype to some readers—somebody's always claiming that *something* is the best new high-performance trick, right?—but in the case of superchargers you will get little argument from engineers, engine builders, or race car drivers. The supercharger's effectiveness has been known almost since the engine was invented.

If this is true, some skeptics may ask, then why don't the Detroit auto manufacturers put blowers on all cars? It is simply a matter of economics. Other, less efficient methods of boosting engine output, such as increasing displacement, raising the compression ratio, or enlarging carburetor size, cost considerably less on a mass-production basis. And, even though most superchargers have only one or two basic moving parts, Detroit would view them as an extra mechanical component on the engine that would have to be warrantied and serviced.

Despite such bottom-line assembly line thinking, however, a surprising number of production cars have come equipped with superchargers in the past, both in the United States and in Europe (often as a last resort—the one known to produce results—to boost the performance of an undernourished or poorly-designed engine to keep up with the competition). And, now that Detroit is finally learning that smaller, more efficient cars are better, more and more manufacturers are beginning to install turbochargers on the little motors to make them act bigger and stronger without using up a whole lot more fuel. Some manufacturers have even been considering and testing various types of belt-driven, positive-displacement blowers recently to boost small car performance and efficiency in the lower rpm street driving ranges.

And, by all means do not forget that the ubiquitous GMC supercharger, the big, impressive, whining, air-gulping windmill that immediately comes to

While a big blower will make a killer out of a big engine, today's engine builders—and Detroit engineers—are discovering that belt-driven superchargers are a compact, efficient, and practical way to make small engines operate as if they were twice as big. This is a Dyer DMS-1 blower on a Smokey Yunick-built Buick V-6.

People used to think that blowers were temperamental or unreliable on street engines, but don't forget that most types were initially designed to run thousands of miles on fleet vehicles, such as the ubiquitous GMC 71 series, as shown here on a 3-71 diesel.

mind when the term "blower" is mentioned to any hot rodder, is in fact a regular production item that has been installed by General Motors on millions of hard-working, cross-country-hauling diesel trucks, buses, and other heavy equipment for the past thirty years. Although superchargers have almost always been considered an exotic addition to production cars in the past, and are still considered as the ultimate in bolt-on performance for modified street rods or machines today, they can and should be as practical, efficient, and reliable to run every day on the street as any other working component of your engine. Superchargers have not only been used regularly on diesel trucks for years, but also on aircraft engines, diesel trains, stationary engines, boats, and many other applications.

There are only two serious reasons that I can think of why all performance enthusiasts haven't installed superchargers on their engines: (1) ignorance...an ignorance bred and fostered by the Detroit mentality of economics described above; and (2) cost. Both of these reasons can be refuted by the same argument, however. Instead of bolting a blower onto our motor, hot rodding tradition tells us to bore and stroke, port and polish, change the cam, buy a new manifold, add carburetion, install headers, and so on. What many of us fail to realize, or have never been taught, is that a properly installed and tuned supercharger can do the same job as all of these other speed parts, and usually do it better. Although a complete supercharger kit can be very expensive these days, the total cost should look less "exotic" when you compare it to the accumulated prices of the several traditional performance parts or operations it can replace.

In the following pages we will discuss some of the history and theory of supercharger development, but we will be primarily interested in showing you the several varieties of superchargers and complete kits currently available for street performance engines, how they should be installed

If you have any doubts whether blowers work or not, take one look at professional Top Fuel dragsters or Funny Cars. These 470-inch, 2000-plus horsepower wonders make more power per cubic inch than any other type of motor, and they owe most of it to the big, tight, Teflon-lined, magnesium-case 10-71 to 12-71 blowers.

Superchargers come in several sizes, shapes, and designs. The centrifugal Paxton/McCulloch blower works on a completely different principle than the GMC or other Roots types, but it produces the same result—forcing a greater "displacement" of air and fuel into a given size of engine. This blower setup was used as factory equipment on several cars (Ford, Studebaker, Packard, Avanti) in the 50's and 60's.

and maintained, and how the package can be tuned for maximum performance and efficiency. We will also study the types of changes that should be made to a typical street motor (and those that shouldn't) before a supercharger is installed, as well as a rundown on the best ancillary components for both the engine and the blower (drive systems, linkage, ignition systems, camshafts, exhaust system, water injection, and so on). Plus we will look at some of the older blowers which could be adapted to today's (or yesterday's) engines, we will show how to shop for a good used supercharger, and even how to set up a GMC blower for the street (although we recommend that you leave this last job to the professional shops which have the machines and expertise to do it right).

You will find that there are plenty of ways to supercharge your engine, from polished 6-71's with two four barrels that poke through the hood, to quiet and unobtrusive centrifugal or vane types that tuck next to the engine, and from $3000 complete kits to home-built systems pieced together from a couple hundred dollars worth of swap meet bargains. But no matter which approach you take, as long as you follow the several hints and guidelines offered here to get the system matched and tuned properly, you will certainly agree that no other performance trick can compare with supercharging.

WHAT IS A SUPERCHARGER?

If we are going to spend the rest of this book talking about superchargers, we had better start by determining exactly what a supercharger, or "blower," is. In the broadest sense, a supercharger is anything that will force more air, or air/fuel mixture, into the cylinders of an engine than would be drawn into the cylinders naturally by the suction of the pistons during the intake stroke.

At sea level, atmospheric pressure is approximately 14.7 pounds per square inch (psi)—air exerts this much pressure naturally on everything near the surface of the earth (because of the air's weight). When an engine is "normally aspirated" (unsupercharged), it must rely on this pressure to push the air into the carburetor, through the manifold, into the intake port, and then into the cylinder as the piston "opens" the cylinder to maximum volume when it moves downward on the intake stroke.

Because of restrictions and bends along the path, friction, turbulence, incomplete evacuation of exhaust from the cylinder, and several other factors, the normally aspirated cylinder is never able to completely draw in a "charge" of air/fuel mix. Let's say the total volume of the cylinder is 40 cubic inches. If you turned the engine to bottom dead center on the intake stroke, by hand, and left it there, the cylinder would fill with 40 cubic inches of air through the open intake valve. Atmospheric pressure pushes the air in (which is exactly the same thing as saying the downward-moving piston sucks the air in—whenever you create a space that has *nothing* in it...that is, create a vacuum...atmospheric pressure will immediately push air into that space through any available opening. The force pushing the air in will be approximately 14.7 psi at sea level, less at higher altitudes. But when the engine is running at speed, atmospheric pressure is not great enough to push 40 cubic inches of fresh air/fuel mixture into the cylinder before the com-

Engines like these look impressive and run strong, but they can never equal the performance potential of a supercharged motor. What's worse, they cost nearly as much to build as a mild blown engine, but they run very poorly at typical street-driving speeds. Blowers are actually more practical and efficient for the price.

pression stroke begins.

What we are describing is the *volumetric efficiency* of the engine, which is a comparison of the total amount of air/fuel charge a normally-aspirated engine *could* draw in (total cylinder volume), to the amount (volume) that it actually *does* draw in under given operating conditions. Other factors being equal, an engine's power output is directly proportional to its volumetric efficiency, and the vast majority of our traditional hop-up tricks are directed at increasing this figure: carburetion, intake manifolding, porting, polishing, valve size, camshaft design, exhaust scavenging, and so on.

There's no question that a big, polished blower will draw a crowd at the local drive-in on cruise night. The small block in this early Mustang certainly doesn't *need* a 6-71 with two 4-barrels, but such a system can be made very drivable since blower boost can be easily adjusted (note large blower pulley to reduce drive ratio).

But no normally-aspirated engine can attain 100% volumetric efficiency (except in rare cases when a perfectly tuned, ram-induction intake manifold can accelerate the intake charge sufficiently to fill the cylinders at a certain, narrow span in the rpm range). By increasing carburetor and port sizes, streamlining passages, and evacuating exhaust we can reduce losses, and increase the efficiency to a certain point. But as long as we are relying on atmospheric pressure to push the air/fuel charge into the cylinder—if that remains a constant—we can very rarely reach 100% efficiency.

After you have done everything you can to help the engine draw as much air into the cylinders as possible by normal suction, there remains only one way to get more air/fuel charge in—and it's a simple deduction: increase the pressure pushing the air into the motor; *force* the air in; *pump* the air in. This is what a supercharger does.

So in the broadest sense a supercharger is any device or means for increasing the cylinder-filling (volumetric efficiency) in an engine beyond that possible by the normal suction of the pistons; that is, beyond that possible under the force of atmospheric pressure.

Given this broad definition, we could say that any sort of pump, fan, or wind-blowing device which can force a greater volume of air through the intake system, and/or get a greater volume of air/fuel charge into the cylinder, than is possible under atmospheric pressure, is a supercharger. Under this definition we might say that a tuned ram-induction intake manifold, or even a good air scoop on a fast-moving race car, is a sort of supercharging device. Likewise, it might be argued that nitrous oxide injection is a form of supercharging, since the super-cooling of the evaporating nitrous greatly increases the density of the air charge entering the cylinder—which is slightly different from increasing the volume. (Running your car on a cold night, or at low altitude, produces the same effect in smaller proportion). Mickey Thompson once even tried feeding his funny car engine with very highly compressed air carried in on-board tanks—with some success!

But none of these methods is strictly considered supercharging. For the purposes of this book, we will define a supercharger as a mechanical air pump capable of producing a positive boost pressure in the manifold of the engine at some point in its operating range.

WHAT A SUPERCHARGER DOES

Obviously, we have already discussed to some degree what a supercharger does. It pumps air, or air/fuel mixture, into the engine at a rate faster than the engine can draw it in under atmospheric pressure. Therefore, instead of loading the cylinder with a partial air/fuel "charge" before compression and combustion, a blown engine can fill its cylinders with a *super*-charge which can attain or actually exceed 100% volumetric efficiency.

As we stated, the supercharger can therefore do the job of several other traditional performance modifications aimed at increasing the engine's volumetric efficiency. If you are going to add a supercharger to an engine for street use, be it a 400-plus-inch big block, a medium-size small block, or a pint-size four cylinder, you will very likely be able to pump in all the extra performance you need without changing the carburetor (other than tuning), changing the intake manifold (other than to accept the blower), porting or polishing the heads, enlarging the valves, changing the camshaft, raising the compression, or installing headers. If you are building an all-out race motor, or if the intake passages of your engine are peculiarly restrictive, you would want to make sure that no

Any type of car will benefit from a blower. And you don't necessarily have to have it sticking through the hood. This full custom '50 Ford is a real sleeper with a 6-71 blown smallblock Chevy.

One reason that supercharging has become so popular is that, with today's complete street kits, a blower is true bolt-on horsepower. You don't have to change anything else in the engine to get a substantial performance increase. Best of all, as demonstrated by this Dyer 6-71 blown big block Dodge Charger, a moderate package like this is completely street drivable.

one component in the intake or exhaust systems "fought" the supercharger by overly constricting airflow either into or out of the cylinders. But most of our recent and current "smog" motors are ripe for supercharging just the way they are.

The blower will draw (or blow) more air through the carburetor. Since the carburetor meters fuel into the airstream in proportion to the volume of air it "reads" passing through it, it will automatically increase the fuel supply. Since the blower pressurizes the entire intake manifold and port runners, there is no need to streamline, tune, or otherwise enhance intake flow. In fact, the air/fuel mixture no longer really *flows* through the manifold in a blown engine; instead you might think of the manifold as a pressurized reservoir of fuel/air mix ready to burst through the valve as soon as it opens. As a matter of fact some supercharged engines have problems keeping the intake valves shut and require stronger valve springs to do so.

Because the intake charge is so eager to get into the cylinder of a supercharged engine, there is little need to increase valve timing or lift with a special camshaft. Naturally, certain camshaft profiles will work better than others in blown engines, and we will discuss some of these fine points later. But in general, a supercharged motor is far less touchy about camshaft profile than a normally aspirated engine (the worst thing for a blown motor is a super long duration, big overlap "racing" cam, since it will allow much of the blower's forced induction to go right out the exhaust port). With standard cam timing, however, the supercharger will still help to blow residual exhaust gases out of the cylinder as it forces fresh fuel/air mix in to take its place. Therefore exhaust valve size, port size and shape, and tuned headers are much less important on a blown motor. The supercharger will actually help pump the exhaust out of the engine.

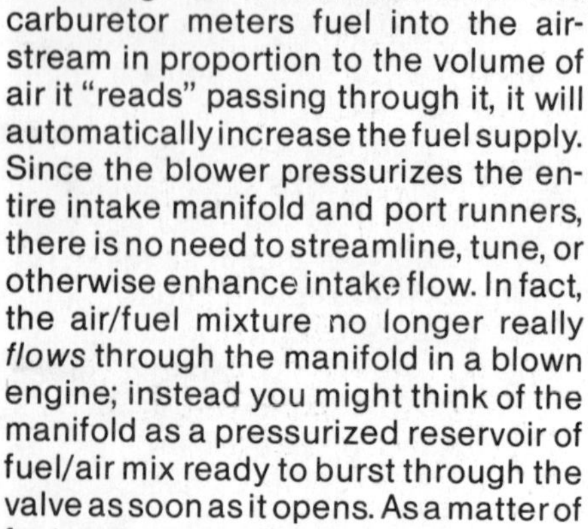

Most people think street supercharging and street blower kits are the latest thing, but they're really nothing new. Back in the early 50's (when most factory engines ran 6:1 or 7:1 compression), a hot horsepower bolt-on was the S.CO.T. blower, which came in kit form to fit flatheads, Cadillacs, Chevy sixes, and other engines.

By pumping a greater amount of air/fuel charge into the cylinder, the supercharger effectively raises the compression ratio of the engine (since a larger volume of air/fuel will be compressed into the same combustion chamber volume at TDC). Consequently the static compression ratio (the physical ratio of the open cylinder volume at BDC to the closed cylinder volume at TDC) in a supercharged engine must be considerably lower than that in a normally aspirated performance engine. Adding a blower to a 10-to-1 or 12-to-1 motor will literally *blow* the engine because of excessively high cylinder pressures. However, our current "smog" motors come with compression ratios in the 8-to-1 to 9-to-1 range, which is just right for the addition of a supercharger (the only consideration is whether the factory pistons and rings are strong enough to withstand higher pressure and heat levels produced in the cylinder by the blower).

Some readers may wonder at the logic of reducing the static compression ratio of an unblown high performance engine so that you can raise the compression back up to a comparable level with a supercharger. If that's all the blower did (raise the compression), it would be a waste of money (and engine horsepower to turn it), since a set of high-compression pistons costs far less than a blower. But don't forget that the normally-aspirated engine—no matter what it's compression ratio—can only draw so much air into the cylinders under atmospheric pressure. It cannot achieve 100% volumetric efficiency. It just squeezes the same volume of air into a smaller space with a taller piston dome. Consequently the higher-compressed charge will fire with a bigger "bang," producing more power than the same charge in a lower compression motor. In a blown motor of the same displacement, however, you start with a larger volume of air/fuel in the cylinder, and then compress it into a larger combustion chamber, arriving at the same proportional compression ratio. Therefore you get the same "bang" as in the high-compression engine, but you have a larger charge to expand and push the piston down. The more fuel and air you put into each cylinder, and burn, during each cycle of the engine, the more power that engine will make. This is what a supercharger does.

Using this same reasoning, you can also think of the supercharger as a device which effectively increases the displacement of an engine. That is, if you compare an unblown 450 cubic inch engine with a supercharged 300 cubic inch engine, you would very likely find that the smaller engine is pumping the same mass of air/fuel charge into its cylinders as the larger engine can draw in naturally. The power produced by an engine is directly proportional to the mass (or weight) of air/fuel mixture it can ingest, compress, and ignite per cycle, other things being equal. That is why larger displacement engines make more power than smaller ones. But rather than making the cylinders bigger, we can use a supercharger to pump more air/fuel charge into the smaller motor, thereby making it per-

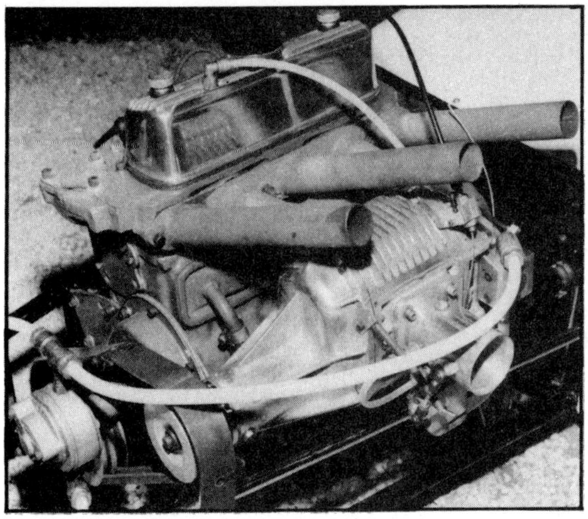

Europeans have driven small cars with small engines for years, and they discovered long ago that superchargers—especially Roots blowers—are the most effective means for making pint-sized motors act a lot bigger, as demonstrated on a pair of Morris fours, one for street, one for drags. American manufacturers (and drivers) are now learning the same thing about small cars.

form just like an engine of much larger displacement. Better yet, the smaller displacement motor can be smaller externally, and lighter, thereby allowing for a smaller and lighter vehicle package, which will increase performance, economy, and overall efficiency tremendously. This is exactly the kind of performance we are looking for today, and superchargers will certainly play a growing role in the small-car era.

Certain types of superchargers provide further benefits. For instance, the Roots blower, which is always used in a draw-through configuration (the blower is placed between the carburetor and the intake manifold), helps to atomize the fuel particles in the air/fuel charge by "mixing" it as it passes through the supercharger. This promotes even and complete combustion in the chamber, not to mention cleaner burning in terms of emissions.

However, we must be fair and point out that not all contributions made by a supercharger are positive. To begin with, all belt-driven superchargers necessarily consume a certain portion of the engine's power, since they are driven by the crankshaft. At high boost levels, this power drain can be considerable—*but the extra power produced by the blower more than offsets this drain.* A good supercharger gives much more than it takes. Turbocharger devotees point out that turbos deliver "free" horsepower, unlike belt-driven blowers, but the truth is that turbos do cost something in backpressure. A Roots type blower, on the other hand, will actually "windmill" under high vacuum driving conditions, such as constant speed cruising. In other words, when the engine is running with the throttle closed or only slightly open, vacuum rather than boost will be produced in the intake manifold, and this vacuum will turn the supercharger rotors like a windmill—and under such circumstances the blower will not be using up crankshaft horsepower. To prove this point, you can disconnect the drive belt from a Roots blower and watch the drive pulley spin by itself while the engine is running.

The only real problem with street supercharging is that it's easy to get carried away. The temptation is always there to dial in a little more boost—to go for more horsepower. The alcohol-injected rat motor in this Corvette is a goer and an attention-getter, but it's not exactly what we mean by a "street supercharged" motor.

Finally, all superchargers raise the temperature of the intake charge as they pressurize it, because the temperature of any gas rises when it is compressed. The amount of intake charge temperature increase will vary with the amount of boost being produced and the efficiency of the supercharger (which we will discuss more fully later), but this increase in intake temperature can lead to detonation in the combustion chamber, and possible parts damage, if not controlled.

TYPES OF SUPERCHARGERS

There are any number of ways to pump things. Just think of the numerous types of water pumps or air fans you have seen. Most of the designs that have been used for superchargers over the years—including current styles—were actually intended for some other purpose originally. You have probably heard the story that the Roots supercharger was initially used to pump fresh air into mine shafts. Actually, it was first used to blow air into forges—as were several varieties of centrifugal "blowers"—well before the internal combustion engine was ever invented. Roots blowers have also been used for such varied purposes as pressurizing airplane cabins and pumping flour or molasses, just to name a couple.

You have undoubtedly also heard the expression, "the internal combustion engine is basically an air pump." Certainly it is—a piston-and-valve arrangement being a common air pump-

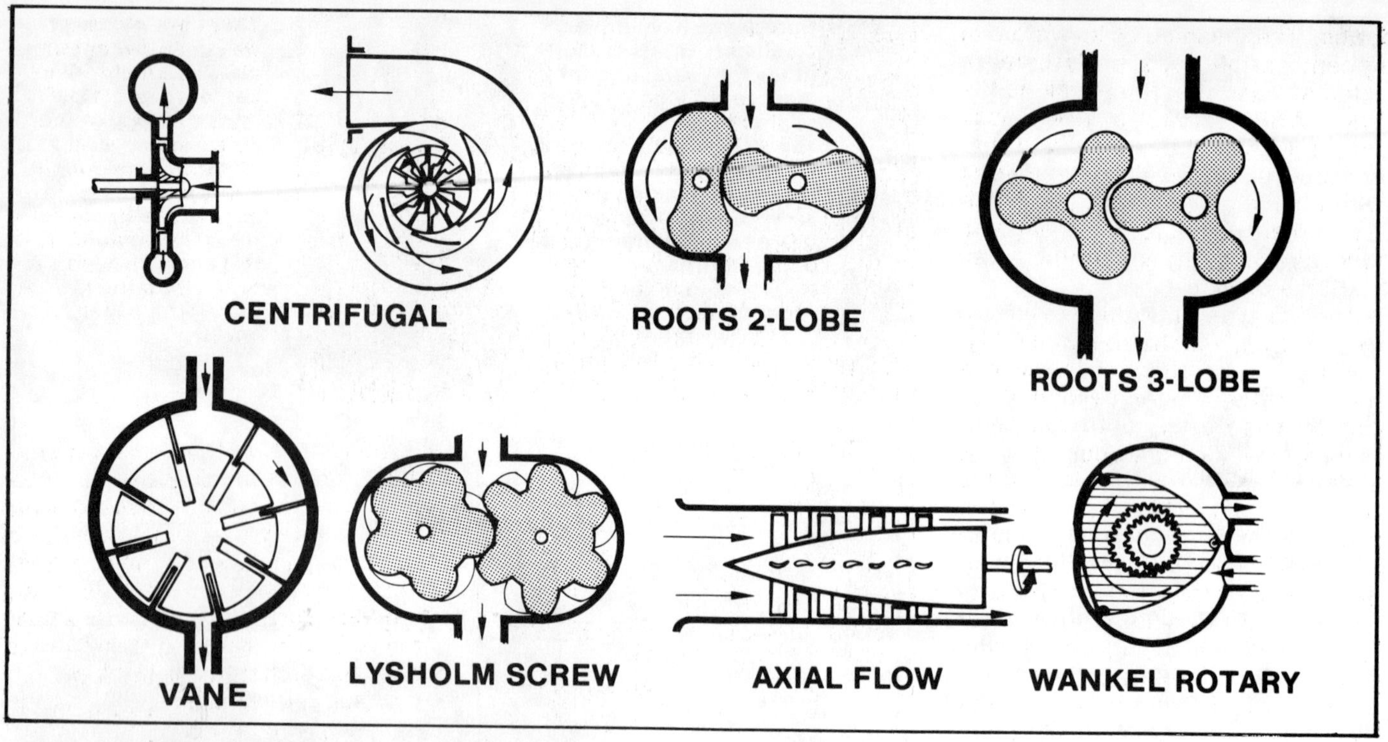

Blowers may look mysterious and complicated from the outside, but most—especially GMC's—are really quite simple machines. This drag-style street roadster runs a 4-port injector atop a 6-71...not very drivable, but a crowd-pleaser.

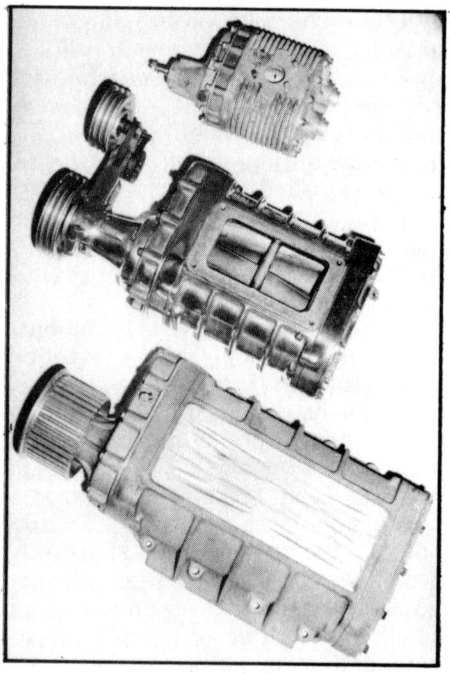

Roots type blowers come in all sizes. There are several sizes and styles of the familiar GMC, plus numerous smaller varieties. Shown here for comparison are a 6-71 GMC, a 4-71 GMC, and a small early S.CO.T. blower. Each pumps a given volume of air per revolution, roughly proportional to the size of the blower.

ing device (most air compressors are of this type). Therefore it is not unthinkable to use a couple of the cylinders in an engine to supercharge the other cylinders, and a few early examples of such "self-supercharging" engines were actually designed and tested, but both the mechanical efficiency and the pumping rate of such a design were too low to be practical. A piston-and-valve air pump is good for high-pressure/low-volume work (such as an air compressor, which uses a large storage tank), while an engine supercharger needs to pump a large volume of air at a relatively low pressure.

Of course there are other types of internal combustion engines besides the piston variety—which are also types of air pumps—and some of these have been used as superchargers. The Wankel "rotary" engine, for instance, was first designed as a supercharger and was used very successfully on some record-setting NSU motorcycles years ago. The jet engine, or "radial fan," is also a type of blower, and this same principle was used in the Latham "axial flow" supercharger made in the 50's and 60's.

Countless other varieties of pumps, compressors, blowers, and variations on each have existed in the past, and may well appear in the future. But at present three major types of blowers are the most commonly used designs for automotive supercharging, and will be the ones discussed in greater detail in subsequent chapters.

The type of supercharger most readily associated with hot rod engines is the Roots type. GMC blowers are of this design, which may be described as a bi-rotor, intermeshing gear pump. A pair of rotors turn inside a closed housing (with an intake opening on one side and an outlet opening on the other) in opposite directions, meshing like a pair of gears. Most Roots blowers have two or three lobes per rotor (though they could have more). The open space between the lobes "catches" the air at the intake opening, carries it around as it rotates (sealed in by the wall of the case), and then "pushes" it out the opening at the bottom. The meshing of the rotors in the center of the blower seals that area so the air cannot escape back up the middle. A Roots blower is called a positive displacement pump since it theoretically moves the same volume of air through it on each revolution. The oil pump in most automotive engines works on exactly the same principle as the Roots blower.

Another type of positive displacement air pump that has been used for automotive supercharging is the vane type. There have been several variations on this design, but basically it consists of a cylindrical housing with a single cylindrical "rotor" which rotates on an offset axis. This rotor is slotted for any number of protruding "vanes" (usually two to four) which maintain light or near contact with the walls of the case at all times. Thus, as the rotor spins, the vanes sweep air into the pump through the inlet port, pushing it around between the rotor and the case. At the same time, as the offset rotor moves closer to the case while it spins, it compresses the air trapped between the vanes before pumping it out the outlet port. Several types of vane type superchargers were made during the 20's and 30's, but they haven't been very common lately (the Judson and the Shorrock being two of the more recent, besides a Bendix blower that never reached production). The Wankel design is a variation on the vane type of pump.

The third common type of supercharger is known as the centrifugal blower. It moves air by means of a bladed fan which spins at high speed inside a spiral housing shaped sort of like a snail's shell. Air enters at the center of the fan (called an impeller), is picked up by the spinning blades, and is then rapidly accelerated by centrifugal force before being flung out the outlet port. The centrifugal supercharger is not a positive displacement type, the delivery air volume increasing as the square of the rotational speed of the impeller. Consequently centrifugal superchargers must be turned at extremely high speeds to produce usable boost, and boost drops off significantly at lower engine speeds. Common centrifugal blowers were the types used on American high performance cars of the 20's and 30's such as the Duesenberg, Cord, and Graham. The McCulloch centrifugal blower, installed on Fords, Studebakers, and Packards of the mid-50's, survives as the Paxton supercharger today. And a new centrifugal blower has just recently been introduced by a company called Power Concepts in California.

There are plenty of ways to blow or pump air. The lobster-like Latham "axial flow" supercharger of the 60's uses several sets of radial fans (much like a jet engine) to accelerate air/fuel mix into the manifold. This is a draw-through system, since the blower pulls air through the carbs.

Centrifugal blowers, like this Mach 2 installed on a Jeep, use a fan inside a snail-like housing to blow air into the motor. This one is a blow-through system, since the supercharger blows air into the carburetor, which is mounted in the normal position on the engine.

Of course, all of the common types of exhaust-driven turbochargers, which have received so much attention from racers and auto manufacturers alike in the past few years, are centrifugal superchargers. The only difference is that they are operated by a turbine driven by the engine's exhaust, rather than by a belt connected to the engine's crankshaft. Although we may make a few comparisons between belt-driven superchargers and turbochargers later on, we will not discuss turbos in this book. The subject has been covered in great detail elsewhere, and certainly deserves separate study. However, I would point out that the intense recent interest in turbocharging has produced significant technical gains in that field of late, many of which can be—or hopefully soon will be—carried over into belt-driven supercharger technology.

In the meantime, however, supercharger design, theory, and application as it is practiced today by most hot rodders, drag racers, and other high performance enthusiasts has actually developed very little since World War II (or even before). Most of us are still simply amazed at the huge amounts of horsepower that can be pumped into an engine by a 6-71 Jimmy. Superchargers are relatively simple devices, to be sure, and we should keep them that way. But today's blower designs are still plenty short on efficiency (well, so are today's gasoline engines...that's nothing new). However, it looks like Detroit's and Europe's current infatuation with the turbocharger might be in the initial stages of shifting toward belt-driven, positive-displacement superchargers as a more practical and effective method of increasing overall power and efficiency—especially of our current smaller engines—in the street driving rpm ranges. If this is true, look for some dramatic developments, or at least some long-awaited refinements, in supercharging in the years ahead.

When GMC blower rebuilders started making manifolds and drive kits—with provisions for street accessories—to fit a wide variety of engines, the street blower movement really took off. This is a lineup of Don Hampton 6-71's for Chevies, Fords, and Mopars.

CHAPTER 2
BLOWER BACKGROUND

Supercharger development in the United States centered around centrifugal blowers. Blown Deusenberg and Miller straight eights totally dominated the Indy 500 through the 1920's. This is a Goosen-designed, Offenhauser-built Miller 91 circa 1927 showing the flywheel-driven centrifugal blower at the back with a single updraft Miller barrel valve carb. Note increasing diameter of the blower scroll housing.

A SHORT HISTORY OF AUTOMOTIVE SUPERCHARGING

Most types of superchargers, in elementary forms, were used for some other purpose well before cars were invented—to pump water, to be pumped *by* water (that is, to work as a type of water wheel or mill), or to blow air for ventilation, for furnaces, or other purposes. The Roots blower design—the type used by GMC and several other automotive supercharger manufacturers—traces its origin to two American brothers, Francis and Philander Roots, who were trying to design a better water wheel to power their woolen mill in 1854. Their bi-rotor gear pump was not successful as a water mill, but they later found it to be very good for pumping large volumes of air at relatively slow rotor speeds as a blast blower for a foundry furnace. The Roots brothers first patented their blower in 1860, and updated the design with 16 more patents by 1884. The Roots blower has been used in a wide variety of industrial applications—including blowing fresh air into mine shafts—ever since, and the company is still operating as the Roots Division of Dresser, Inc. of Connersville, Indiana.

The first noted example of a blower being fitted to an internal combustion engine was in 1901 when Sir Dugald Clerk, the pioneer of two-stroke engine design, used a pumping device to force more air into the engine to (supposedly) lower the maximum combustion temperature. As we shall see, blowers actually raise the temperature of inlet air and of combustion in the cylinder; but Clerk found that the blower ultimately increased the power of his two-stroke engine by six per cent. Thus the power-producing capabilities of supercharging were discovered. In 1902 Louis Renault patented a supercharging system in France which used a centrifugal compressor to blow air into the carburetor.

The first example of supercharging in the United States, the first supercharged car to win a competitive event, and the first application of a supercharger for production vehicles all fall to Lee Chadwick of Pottstown, Pennsylvania, who had been manufacturing Chadwick automobiles since about 1900. In 1907 he and engineer John Nichols mounted a centrifugal blower to the 1140 cubic inch Great Chadwick Six, producing a certified top speed of over 100 mph. The initial Chadwick design drove the supercharger with a leather belt from the flywheel at nine times crankshaft speed, blowing dry air into the carburetor. A second version used a 3-stage centrifugal compressor with three 10-inch diameter impellers of twelve blades each (but of increasing widths for 3-stage compression). This car won the Wilkes Barre Hill Climb in 1908, the first event in which a supercharged car was ever entered, and went on to score many other victories before Chadwick ceased production in 1911 after making some 260 cars. The Chadwick was obviously also our first street supercharged vehicle.

Supposedly Chadwick had considered an exhaust-driven centrifugal supercharger (that is, a turbocharger) before deciding on the belt-driven type. However, a Swiss engineer named Buchi is generally credited with the invention of the turbocharger. He adapted an exhaust-driven blower to a diesel engine in 1909.

In 1911 a Frenchman named Sizaire carried out experiments with centrifugal blowers, and the following year Marc Birkigt designed a Hispano-Suiza for the Coupe de l'Auto race which used two of its six cylinders to supercharge the other four (that is, a four cylinder engine with a built-in two cylinder compressor for supercharging). This experiment was not very successful.

Further automotive applications for superchargers were eclipsed by the outbreak of World War I. This was the first air war, and air superiority relied heavily on supercharging, because the higher an airplane flies, the less dense is the surrounding atmosphere to feed the engine. Supercharging was immediately seen by the Germans as a practical means for maintaining engine power at higher altitudes, by mechanically forcing a denser charge into the cylinders. Similar supercharger development followed in other European countries and in the U.S.

Naturally, most leading auto designers and manufacturers were involved in the war effort in Germany and elsewhere. Not long after the war's end superchargers of various types began showing up on race cars.

One of the first companies to use superchargers for automobiles in Germany was Daimler (which made the Mercedes cars, later to merge with Benz in 1926). With the assistance of Dr. Porsche, who worked for Daimler at the time, they developed the 28/95 Mercedes 6-cylinder which won the 1921 Coppa Florio—the first victory for a supercharged car since the Chadwick. The 28/95 Mercedes, like several later models including the famous SS and SSK's, used a Roots-type blower mounted vertically at the front of the engine, driven by a bevel gear and cone clutch from the crankshaft. The blower was engaged only at full throttle, and blew into the carburetor. When the blower was engaged, power was reportedly increased 50% with a pressure boost of approximately 6-7 psi. A considerable amount of Roots blower technology was compiled by Mercedes Benz between WW I and WW II.

In 1923 Fiat added a Wittig vane-type supercharger to their 8-cylinder Gran Prix car. The blower was front-driven by the crankshaft, and, like the Mercedes, blew into the carburetor only at full throttle. This early vane-type blower did not work very well (possibly because oil had to be injected into the blower to lubricate the vanes, causing detonation in the engine). Later the same year Fiat switched to a Roots blower with an intercooler and did much better, finishing one-two in

One of the more radical Indy designs was the Deusenberg "side-winder" introduced in 1927, using a side-mounted, shaft-driven centrifugal blower. All of the fins on the blower case and the tall intake manifold weren't just for looks. Their purpose was to dissipate heat generated by the blower—one of the first intercoolers.

European manufacturers preferred positive-displacement, Roots-type blowers. One of the more famous early blown cars was the Mercedes SSK (this is a '29 model), which used an upright Roots-type supercharger driven by a bevel gear from the crank, with a clutch to engage the blower only at full throttle. Although this car won many races, it was outfitted and sold as a street machine, as were most of the early supercharged "sports cars."

Another famous supercharged race car/street machine was England's Blower Bentley. As shown on this partially disassembled '31, the big Roots-type blower is crank-driven and draws through two SU carbs. Looks just like the front-drive GMC set-up popular on dragsters of the early 60's, doesn't it?

the Italian Gran Prix and setting a new lap record just shy of 100 mph at Monza.

In 1924 both Sunbeam and Alfa Romeo introduced Roots-blown Gran Prix cars, with the Sunbeam being the first to place the Roots supercharger between the carburetor and the intake manifold (that is, in a draw-through rather than a blow-through arrangement), and discovered through testing that this design netted a 20% increase in horsepower over the blow-through configuration. The cooling effect of the vaporizing fuel in the blower, plus the better atomization of the fuel particles as they passed through the rotors, were the main factors in the power increase. After this, until a rules change in 1952 that virtually banned blowers, supercharged cars totally dominated Gran Prix racing in Europe, and the majority of these cars used Roots-type blowers.

In the United States, by a series of coincidences, supercharger development centered around the city of Indianapolis. Prior to WW I, Dr. Sanford Moss, working for the General Electric company, had been experimenting with Roots-type blowers and centrifugal turbochargers; he was also a pioneer of the 2-stage blower concept. Dr. Moss, as well as Kerr and Le Blanc who were designing a gear-driven centrifugal blower for the Liberty aircraft engine, were working for the U.S. government at McCook Field, near Indianapolis. Allison Engineering, manufacturer of the Liberty engine and other subsequent supercharged aircraft engines, was also located near Indianapolis, and the company's owner, James A. Allison, was one of the founding owners of the Indianapolis Motor Speedway. Thus much of the pioneering supercharger development in the United States was occurring in the same city as most of our pioneering race car development, albiet for different reasons.

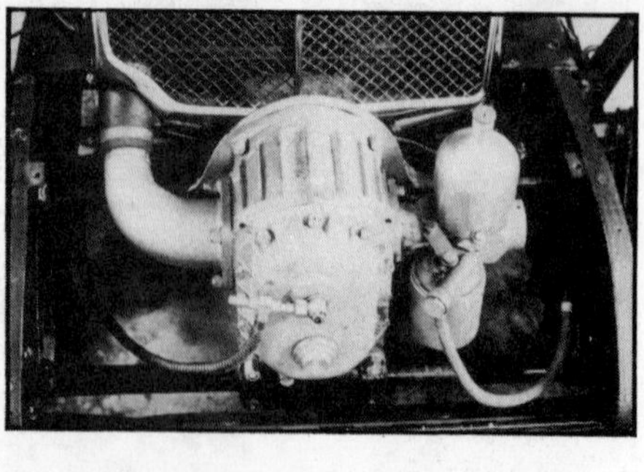

The 1934 MG K-3 "Brooklands" also used a front-drive Roots type blower (a Marshall) with a single SU carb.

Meanwhile brothers Fred and August Duesenberg had opened their new automobile manufacturing company in Indianapolis in 1912, and were involved in building one-off race cars. In 1924 the Duesenbergs fitted a centrifugal supercharger to their already highly successful straight eight racing engine. The blower was gear-and-shaft-driven, and was the first recorded example in the U.S. to place the carburetor ahead of the supercharger, in a draw-through layout. This supercharged Duesy was a winner, and all subsequent Duesenberg racing engines were similarly supercharged. After Errett Lobban Cord bought into the Duesenberg company in 1928, as he had with Auburn previously, they introduced the famous J and SJ production models, which came equipped with shaft-driven centrifugal superchargers mounted horizontally above the engine. Models of the Auburn, the V-8 Cord, and the Cord look-alike Graham also came factory-equipped with similar centrifugal blowers.

The Italians were also big on Roots blowers, and most manufacturers made their own. These two straight eight Bugattis use similar side-mounted blowers driven by a shaft from the crank gear. An updraft carburetor mounts under the blower. The single overhead cam is a '27 Type 35B, the DOHC is a '34 Type 55.

Shortly after the Duesenbergs unveiled their blown engine at the 500, another Indianapolis-based engine builder and a fierce competitor of the Duesenbergs, Harry Miller, came up with a centrifugal supercharger for his straight eight racing engines. Designed by Leo Goosen and built by Miller machinist Fred Offenhauser, this flywheel-driven blower was relatively advanced for the time (more advanced, in fact, in terms of impeller, diffuser, and scroll-housing design, than some of our current offerings in aftermarket belt-driven centrifugal blowers).

The most interesting point about this period of supercharger activity in motorsports (roughly between WW I and WW II) is the fact that Roots-type blowers were used almost exclusively in Europe while centrifugal-type blowers were used exclusively in the United States. The Europeans tried other types of blowers to begin with, as did the Americans, but each group soon discovered which type of supercharger worked the best for the particular style of racing predominant in that part of the world. In Europe the Gran Prix, or road racing, was the ultimate in auto competition. The twisting and turning road course required a broad horsepower range in the engine, as well as instant throttle response and good low-end torque for accelerating from slow corners. In the United States, however, the ultimate was the Indy 500—a long, fast oval with sweeping curves that kept a racing engine operating in a fairly tight rpm range. From the turn of the century on, auto racing in the U.S. was done primarily on circle, oval, or board tracks, and therefore mostly at wide-open throttle. The Roots blower gave the Europeans the kind of power they needed; the centrifugal blower, which was more efficient in a narrower rpm

On the 1934 Alfa Romeo Tipo 8 "Mille Miglia" the Roots-type blower tucks low on the side of the engine with a big Memini carb just ahead of it. Again, note the use of fins on the manifold for intercooling. This car won Le Mans four times in the 30's.

In the 20's and 30's you could buy several types of supercharged production cars to drive on the street (if you had enough money). One of the most famous in America was the Deusenberg SJ, which used a horizontal, shaft-driven centrifugal blower. This example, a '35 SSJ owned by Gary Cooper, was one of three specially outfitted with two updraft carbs by the factory.

range, gave the American cars the top-end boost they required.

The period between the two World Wars was the heyday of supercharging. Not only did the major race cars of the time run blowers, but as you might expect, so did several of the more racy production models, both here and in Europe. However the economic Depression and then WW II finally put an end to supercharged luxury sports cars, and shortly after war's end rules changes stifled the supercharger's advantage in racing. Consequently, after the late 1940's superchargers virtually disappeared from the automotive scene.

Just before and after World War II was when hot rodding really began to develop, however. The young kids who raced stripped-down Model T's and A's on dirt ovals and dry lake beds were a far cry from the sophisticated European Gran Prix and U.S. Indy race car builders, but they had paid close attention to the way the experts built racing engines, and knew the effectiveness of superchargers. In hot rodding there was no ban on blowers; the major deterrent then, even more so than now, was cost.

Yet there were a few early examples of blowers being run on roadsters at the dry lakes, even in the Depression days before the war. A few rodders acquired centrifugal blowers from production American cars such as the Graham, and adapted them to the Ford flathead V-8. However, Graham production blowers produced very little boost; when hot rodders tried to spin them faster to get more pressure out of them, they usually ended up stripping the blower's step-up drive gears. McCulloch also introduced a centrifugal supercharger as an aftermarket bolt-on kit for the Ford

Pre-war supercharging reached its peak in Europe with the two-stage, Roots-blown Mercedes W163 Gran Prix cars of '39. These blowers were crank-driven at the front of the engine, with the larger blower pumping into the smaller one to step up pressure to an amazing 26 pounds of boost, producing 483 h.p. from just 183 cubic inches! Technical features of this set-up abounded, including full-throttle fuel enrichment. The sophistication of today's typical street blowers pales by comparison.

During World War II supercharger development progressed rapidly, but for airplane engines rather than for automobiles. The huge unit at the front of this Rolls Royce Merlin V-12 is a two-stage, crank-driven, centrifugal blower. Unfortunately, little of this technology has ever been applied to cars.

One of the first Roots blowers to be used on a hot rod was a Mercedes unit which Don Blair ran on his Modified lakes roadster in 1946 to turn 141 mph with a Ford flathead V-8.

Barney Navarro pioneered the use of the GMC 71-series diesel blower on hot rod engines after the war. Using a modified 3-71 on a 176-inch flathead V-8, this roadster turned 146 mph at the lakes, plus raced on circle tracks and briefly at the first Santa Ana drags.

V-8 about this time (probably the first aftermarket blower kit), and some of the more affluent rod builders tried them at the dry lakes.

The first generally noted example of a Roots-type blower being fitted to a hot rod was in the late 30's when the Spalding Brothers somehow acquired a Mercedes Benz blower and put it on their flathead Ford. After the war, Don Blair bought this blower from the Spaldings and adapted it to his "Goat" unlimited class roadster for the 1946-47 seasons. Running alcohol through two Stromberg 48's, and driving the blower with two V-belts, the car turned 141 mph.

The person usually credited with the first installation of a GMC industrial blower on a hot rod is Barney Navarro. Barney had been aware of blowers for some time, but they were scarce and expensive. Then one day in 1948 Kong Jackson showed up with a 3-71 GMC off of a WW II landing craft, and offered it to Navarro for $60. Barney was manufacturing his own line of V-8 heads and intake manifolds at the time, so he modified one of his manifold patterns to accept the blower and cast an intake manifold for it. He built his own drive using four V-belts and mounted four Stromberg 48 carbs on top of the blower to run alcohol (ethanol). On a destroked, 176 cubic inch V-8 in his '27 T roadster, this early Jimmy setup pumped 16 pounds of boost and pushed the car to 146.9 mph. The only internal modification Barney made to the blower was to bore the case slightly larger for increased rotor-to-housing clearance for higher rpm use.

Barney not only ran the roadster at the dry lakes, but also raced it on the dirt circle tracks—possibly the only supercharged hot rod to do so. Although he never had any mechanical problems with the blower, he did have plenty of trouble with V-belts, especially on the track, since they tended to heat up and disintegrate. Barney solved the problem by drilling a series of holes both vertically and horizontally through the pulleys to cool the belts.

Early hot rod blower installations consisted of whatever was available. This late-40's lakes roadster uses a Roots blower of undetermined origin on a Cadillac flathead V-8.

Tom Beatty, who worked for Navarro, built a similar system for his own lakes roadster, and then later adapted a 4-71 GMC to an Oldsmobile OHV V-8 for his belly tank Lakester. Beatty eventually manufactured manifolds and drive "kits" for the GMC's, based on the Navarro V-belt design, and helped pioneer the more widespread use of "Jimmies." In fact back in the 1950's Beatty stuffed one of his blown Olds race engines into his Forty Ford sedan delivery to use as a street driver and push truck.

Navarro was also one of the first to run a blower at the drag races, since he entered his roadster at the first running of the Santa Ana drags in 1950. However, Barney did not pursue drag racing with a blown engine because, as he put it, "I broke too many quick change rear ends."

According to hot rod historian Greg Sharp, the first use of a blower at the drags was actually at one of the very first organized drag meets, held near the Goleta air strip in 1949. The contest was between Tom Cobbs, who was running a Roots-blown V-8, and Fran Hernandez, who was using a secret new fuel called nitro-methane.

Hernandez won that round. But it was drag racing that really pushed the development of GMC blowers in hot rodding. The Roots blower produced

Street blowers are certainly nothing new. The year is 1950, and this sanitary flathead-powered street T is running a 3-71 Jimmy adapted with a J.E.M. drive kit.

One of the real innovators in the use of GMC blowers for drag racing was Ernie Hashim. Shown here is his 3-71 blown flathead, built in 1950, and one of the first front-mount set-ups for a Chrysler, which he hand-built.

gobs of horsepower and torque in the low rpm ranges, which is exactly what dragsters needed to get off the line—unlike the Lakes and Bonneville racers who went for top end speed. However the Cobbs/Hernandez duel was also representative of another new element in rodding that affected both drag racing and blower development—nitro. Once the secret of nitro got out, rodders discovered that it gave them even more horsepower than blowers did. But it was also very temperamental, literally explosive, stuff. Early experiments with blowers on alcohol or gasoline proved fairly reliable, but the combination of a blower with nitro usually ended in disaster. So widespread use of Roots blowers in drag racing had to wait until the racers got the nitro situation sorted out.

This is not to say that some drag racers weren't using Jimmies right away. One of the first was Ernie Hashim of Bakersfield, California, who mounted a 3-71 GMC on top of a Ford V-8 as early as 1950 and was using it at the lakes as well as to drag race on airfields in the San Joaquin Valley, such as Maricopa and Famoso. He got his 3-

Chuck Potvin evolved front-drive GMC systems, eventually making good-looking cast aluminum pieces as shown on Ken Lindley's 1959 Reath Auto dragster Chrysler (driven, incidentally, by Don Hampton).

71 from partner John Gilbert, who had operated heavy equipment in the Northwest and had therefore become acquainted with GMC industrial blowers. By 1955 Hashim had adapted a front-mounted 4-71 GMC to a Chrysler hemi with a one-off Hilborn 2-throat injector, which he ran in a '32 roadster at Bonneville and in a rail at the drags. Shortly afterwards Hashim mounted a 6-71 on top of the Chrysler and fabricated one of the first Gilmer belt drives. About the same time Ed Iskenderian purchased a batch of 6-71 GMC superchargers and hired Hashim to rework them for use on dragsters, which involved reversing the rotors, milling the case, and resetting the clearances. In 1959 Isky introduced their "Forced Induction" blower drive kit for top-mounted 6-71's.

Prior to that time the most popular blower set-up for drags or Bonneville was Chuck Potvin's front-drive for 6-71's. Early versions, introduced in the mid-50's, were of a "U-Fab" variety; later models used cast aluminum manifolds and drive housings. The big advantage of the front-mount blower, besides the fact that the lower profile was better aerodynamically and gave the driver a better view, was that the direct drive from the crankshaft solved the persistent problem of drive belt slippage when the blower was pumping substantial boost (the toothed Gilmer belt would later solve this problem for top-mounted blowers).

Mickey Thompson probably gets credit for the first blown Chrysler hemi. In 1953 he dropped one of the new hemi-head motors into Bill Burke's old streamlined Bantam coupe—along with one of his more traditional 3-carb Ford flathead V-8's! To this strange marriage Thompson further added a 4-71 Jimmy atop the Chrysler. The first year the blower resulted in two melted pistons in the hemi; later the odd-coupled contraption went well over 190 mph. In 1954 Bruce Crower brought a '49 Hudson to Bonneville, powered by a blown Chrysler, ran 157 mph, and drove it on the street. In '55 Ernie Hashim and Tony Waters set the fastest speed ever for a stock bodied roadster with their blown hemi '32 at 189 mph. That same year Tom Cobbs switched the home-made front-mount 4-71 from his flathead in his '34 coupe to a brand new Chevy 265 V-8, probably the first Jimmy-blown Chevy small block.

Also, in 1955 top dragster pilot Calvin Rice was getting a home-made front drive GMC blown Chrysler hemi to run faster than his back-up flathead in his Nationals-winning rail. However, the 6-71 blown hemi did not become the standard dragster powerplant until about 1959. That was the year of the first Bakersfield Fuel and Gas Championships, and the first trip to the west coast for Don Garlits. When Garlits arrived he was running his fueler naturally aspirated on eight carburetors, but all of his California competitors had recently begun running the big GMC blowers. After that meet, but before his other two scheduled west coast appearances, Garlits towed the dragster immediately down to Isky's to have a 6-71 installed between the hemi and the eight carbs. Other than developments in fuel injection systems, alloy blocks, and big-displacement blower cases (8-71's to 12-71's), the story in fuel drag racing has been the same ever since.

As far as supercharging for street-driven, gasoline-burning automobiles goes, the history is brief. After the prestigious supercharged road cars of the

This unidentified photo was taken at the first World Series of Drag Racing in 1954. The GMC-blown Cad flathead from Illinois was turning a respectable 110 mph.

Can you guess who's driving this GMC-blown, GMC-six-powered drag roadster at Santa Ana in the early 50's? None other than "Dyno" Don Nicholson, who cut his teeth on Jimmies. Note chain drive on side-mounted 4-71.

20's and 30's, the next appearance of blowers was a handful of bolt-on supercharger kits which suddenly appeared on the market about 1950. These included the McCulloch, the Frenzel, the S.CO.T./Italmeccanica, a few kits based on reworked GMC 3-71's or 4-71's such as the J.E.M. and the Speed-O-Motive, and a bit later the Judson and Latham. We will discuss these blowers in more detail later, but they were used primarily in street applications, rather than racing. Because such blower kits were very expensive for the time, not many found their way onto hot rods when they were first introduced.

The next development came in the mid-Fifties. After Chevrolet introduced their new small block V-8 in 1955, Ford and other manufacturers suddenly found themselves in the position of playing catch up. Since Ford's new Y-block was no match for the Chevy, Ford turned to supercharging as the one sure way to best the Chevies on the NASCAR tracks and on the beach at Daytona. They had McCulloch design a special centrifugal blower for the racing Fords, and then offered the standard McCulloch variable-speed belt driven blower as optional equipment on '57 Fords and T-Birds. The same year Studebaker offered the McCulloch on their Golden Hawks and on some of the last Packard models. As late as 1963 the McCulloch/Paxton was used as standard equipment on the Studebaker Avanti. From this time on, however, Detroit turned to the turbocharger for supercharging production models.

Through the 60's and early 70's there were isolated examples of supercharged street-driven hot rods, but most of these were considered phenomenal at the time. The sight and sound of a working GMC blower on the street, not to mention the potential power on tap, was awe-inspiring. But GMC blowers were scarce, parts to adapt them to a street engine were nearly non-existent, and getting one to work right in traffic took lots of trial and error.

Within the last few years, however, blowers have come of age for street machines. Rodders have discovered that, with proper clearances, bearings, and drive ratios, there is no reason why a GMC should be any "hassle" on the street. First a variety of companies offered complete drive kits for Jimmies on street motors—kits which incorporated water pumps, fans, alternators, and so on. Then a couple of companies started fabricating or modifying blower manifolds to fit street motors other than smallblock Chevies and Chrysler hemi's. Now you can choose from a variety of complete bolt-on GMC street blower kits, including a clearanced and polished blower, manifold, drive, carbs, and everything else you might need. In addition, several companies have recently introduced new superchargers, with complete installation kits, designed specifically for street performance engines. And finally, with our current trend toward smaller cars and four-cylinder or V-6 engines, supercharging looks more and more practical—to Detroit manufacturers as well as to aftermarket blower designers—as an excellent means to breathe some new life into these undernourished production vehicles.

Oftentimes hot rodders who tried innovations met with derision from others who didn't understand the principles. Ted Rawleigh fed a crank-driven 6-71 into a belt-driven 4-71 on his Olds-powered dragster in the mid-50's and got 670 h.p. However, the car couldn't put the power to the ground, and 2-stage supercharging hasn't been tried since at the drags.

Through the 1950's most blower applications were strictly do-it-yourself, and often the results were less than professional. Need for substantial guard around chain drive was a lesson learned from experience.

The year of the first Bakersfield Fuel and Gas Championships, 1959, was the year that the 6-71 GMC-blown Chrysler hemi came to rule drag racing...and things have changed little since then.

CHAPTER 3
FINDING SWAP MEET GEMS

SWAP MEET GEMS

Over the years a variety of superchargers and kits has been offered for automobiles. The majority of these appeared in the early 1950's, and consequently most of the early kits were made to fit the Ford flathead V-8. With the growing interest in hot rod memorabilia, and the increasing construction of early-style hot rods, these older blowers are becoming collectors' items and sought-after additions for present-day flatheads and other nostalgic motors.

Yet, even though the demand for old superchargers seems to be growing, prices for these relics appear quite reasonable and the supply of various types of old blowers at the swap meets is surprisingly good. Since these superchargers have only a few moving parts, and since the major components (the case and the rotors or impeller) should not wear at all, they can usually be refurbished with new seals and bearings and be expected to function as well as they did initially.

But this isn't saying much. Most of these early blower kits (with the exception of the McCulloch/Paxton and the Latham) were designed to give good bolt-on performance increases to stock, small-displacement, low-compression engines. If you are looking for a practical performance increase on a modern V-8 of 300 cubic inches or more, look at one of the several new superchargers or GMC blower kits detailed later in this book. But if you are playing around with old time speed equipment, building a nostalgia rod, or even looking for moderate street boost on a small four or six cylinder,

Swap meets abound with used blowers and accessories. Some of it is junk, but many used blowers and components are real bargains at swap meet prices.

Typical of swap meet bargains, these two early Paxton blowers were selling for about $200 each, and were in fine condition.

one of these old blowers could be a real bargain at the swap meet, and would certainly be a conversation piece on your motor.

EARLY McCULLOCH BLOWERS

As far as I can determine, the first aftermarket blower kit made in America was a centrifugal supercharger made by Robert Paxton McCulloch in the late 1930's. This initial McCulloch blower was designed to bolt onto the manifold of a Ford V-8 and was driven by a pair of V-belts from the crankshaft. The impeller and scroll housing of the blower mounted horizontally above the intake manifold, the impeller being driven by a bevel gear from the belt-driven shaft. A single Ford 2-barrel carb mounted on top of the blower housing, which was made of aluminum.

This initial McCulloch blower, made in Milwaukee, was similar to the centrifugal superchargers used on Duesenberg SJ's, Auburns, and Grahams. McCulloch was a Stanford engineering graduate who was building racing engines for boats in the Milwaukee area in the 30's. He would later develop superchargers for the army during WW II, and then move to California to establish the McCulloch Corporation which made, among other things, McCulloch/Paxton superchargers, small engines, go-karts, and chain-saws. These early McCulloch blowers reportedly showed up on a few pre-war Lakes racers. The Granatelli brothers were using them on their own hot rods in Chicago as early as 1938, and became a distributor for them (they eventually bought the Paxton blower division from McCulloch in 1957). But very few of these early flathead McCullochs exist today. Consider them strictly a collector's item.

The later McCulloch/Paxton centrifugal supercharger, introduced in 1953, is essentially the same as the Paxton blower being marketed today, and it is probably your best swap meet bargain if you are looking for a good older blower to actually make some boost on a street engine. Early units were made by the Paxton division of McCulloch Corp., but carried a McCulloch emblem on the housing. These belt-driven blowers were sold in complete bolt-on kits to fit most cars of the early to mid-50's including Fords, Cadillacs, Oldsmobiles, Buicks, and so on. The first O.E.M. (original equipment manufacturer) contract for McCulloch was with Kaiser in 1954, and the blowers were installed on all Kaiser Manhattans that year; the next year

The first aftermarket street blower kit was this centrifugal McCulloch marketed in the late 1930's. The horizontal blower is driven by a bevel gear shaft. Note that polished unit on engine differs from earlier cutaway model.

Kaiser went out of business. We will discuss the Paxton supercharger in more detail in the section on centrifugal superchargers—but if you see an older one at a swap meet for a good price, it could be a smart purchase.

THE FRENZEL

Another early centrifugal supercharger, introduced in 1950 as a bolt-on for Ford V-8's, was the Frenzel. Manufactured in Denver, it was similar in design to the initial McCulloch, except that it bolted to the stock intake manifold with the impeller and scroll in a vertical position rather than horizontal, and it accepted two 2-barrel carbs instead of one. A spur gear housing in front of the 7-1/2 inch impeller ran the blower at 5.9 times crank speed. This blower was rated at 6 pounds boost at 4200 rpm on the flathead, which would reportedly increase horsepower 50% over stock.

To give a good idea of how well these early blowers worked on production cars of the time, a stock '47 Ford coupe was fitted with a Frenzel blower and tested at El Mirage dry lake in 1950. Without the blower it ran 84.6 mph top speed and accelerated from zero to 60 mph in 16.3 seconds. With the blower it did 100.05 mph top speed and 0-60 in 11.0 seconds. This was a substantial improvement for a single piece of bolt-on speed equipment, and helped justify the $175 price tag (which was also substantial for 1950). However, Frenzel blowers are extremely rare these days. Apparently very few of them were ever produced.

If you do find one of these early

The mysterious black box on John Bandimere's '55 Chevy from Denver, Colorado, is a unique, one-off McCulloch Roots-type blower. With it, John ran sizzling times at the '55 National drags in Great Bend, amazing the crowd. Bandimere learned early about blowers' effect in mile-high Denver, and was a supercharging expert in the early 50's.

Another early one-off McCulloch (Paxton) blower is this centrifugal unit made for extremely high boost on the Novi Indy cars. Note similarity in design to Miller Indy blower shown earlier.

centrifugal blowers at a swap meet (including a Graham, Auburn, or Duesenberg production blower), the first thing to check is the condition of the step-up gears which drive the impeller. Although belt slippage would probably save the gears on a McCulloch or Frenzel, the extreme inertial loading from the high rotational speed of the large-diameter impeller tended to strip the drive gears as the engine was revved rapidly up or down. Second, check the condition of bearings inside the blower, which also tended to succumb from the high rotational speeds (the Frenzel was lubricated with engine crankcase oil, which could be detrimental if the owner neglected oil changes). Worn bearings and seals could very likely be replaced with modern types available in a huge variety of I.D. and O.D. sizes at bearing supply houses. Worn or damaged drive gears would be much more difficult to replace.

THE S.CO.T/ITALMECCANICA

By far the most popular, and apparently the most plentiful, of early blowers is the Italmeccanica/S.CO.T. This small, good-looking, dual-lobe, Roots-type supercharger was first produced in Italy as the Italmeccanica and was imported as late as 1951 in complete kits for American cars such as Ford and Merc flathead V-8's, Chevy 6's, Crosleys, and Studebakers. These blowers were made in different sizes for different engines (they were also made to fit several European cars as well), but the most popular unit was that for the flathead V-8. The kit included a new manifold to accept the blower, with a pop-off valve, plus all necessary pulleys, generator relocation mount, and belts. The top of the V-8 blower case accepted two Ford 2-barrel carbs. The "I.T." blower, as it was often called, produced 6-8 pounds boost and increased horsepower on a stock 59A Ford V-8 from 82 to 150 at 4000 rpm in one reported dyno test.

Later, American versions of this blower were manufactured in New Jersey (using American thread and bearing sizes) and were called the S.CO.T. (probably for the Supercharger Co. of Turin). Larger versions of the S.CO.T. were subsequently made for Cadillac and Oldsmobile OHV V-8's in the early 50's. However, as of 1952, a flathead V-8 S.CO.T. kit listed for $467.50, and the Cad/Olds kits went for close to $800. That's considerably more than a whole engine cost back then. Today, if you're lucky, you can pick up decent S.CO.T. blowers for well under 1952 prices.

Although you may have to make your own mounting brackets and belt drive, you can easily adapt early McCulloch/Paxton superchargers to most any engine since they blow into the carburetor(s) on a standard manifold. This one is hooked up to a dual 4-barrel tunnel ram on a big block Chevy, and has seen thousands of street miles.

EARLY GMC KITS

Of course, as we have already discussed, rodders had also begun to discover the GMC Roots-type blower by the late 1940's. One of the first to offer automotive adaptation kits for 3-71 and 4-71 Jimmies was Tom Beatty, first for flathead V-8's and later for Oldsmobile overheads (Beatty made an intake manifold for the Olds which has his name on it). Chuck Potvin made front-drive GMC kits beginning in the mid-50's, and Isky finally came out with a complete top-mount 6-71 drive and

An early photo shows a Frenzel blower being tested on a "back yard" dyno on a healthy flathead.

The only example of a Frenzel blower I've ever seen was this complete unit offered at a swap meet in Maryland over ten years ago.

The only difference between Italmeccanica (right) and S.CO.T. blowers of the early 50's was that the S.CO.T. had American bolts and threads and was imported from Italy in kits to fit U.S. cars. Both blowers, made in Turin, came in several sizes.

manifold kit for Chryslers and Chevies in the late 50's. However, these GMC blower kits were all designed for competition, not for street use.

In about 1951 or 1952 both Speed-O-Motive and a short-lived company called J.E.M. (Jack and Ernie McAfee) tooled up intake manifolds, blower covers and snouts, and V-belt drives to mount 3-71 and 4-71 GMC's on Cadillacs and Ford flatheads, and possibly other popular engines. These parts are quite rare.

The first really complete and widely marketed kit for adapting a 3-71 or 4-71 GMC to a street engine was Cragar's belt-drive set-up introduced in about 1959. This kit included an intake manifold (which would accept either a 3-71 or 4-71) with a stock-type water neck, a drive snout and shaft for the front of the blower, a thin rear cover, a 4-barrel carb intake, and a triple V-belt drive system. You had to supply your own blower, carburetor, and linkage. This kit was intended primarily for street applications, and tons of them were sold, mostly for small block Chevy V-8's (although advertising states that they were available for "most popular V-8's"). Andy Brizio, owner of Champion Speed Shop and builder of Andy's Instant T roadsters, bought this kit from Cragar in the late 1960's, and marketed it for a time with the name "Andy's" embossed on the manifold (kits were available for small block Chevies and Fords). These kits are now made and marketed by Cub Barnett in Belmont, California.

After drag racers began using GMC blowers in great numbers, numerous speed parts manufacturers, racers, and machinists produced competition-style blower drives, cases, end plates, manifolds, and various accessories over the years. At swap meets or on speed shop used parts racks, you may find blower parts and pieces with names like Pete Robinson, Mickey Thompson, Delta, Van Luven, Danekas, and numerous others—all of which have come and gone in the blower business.

S.CO.T. blowers are surprisingly plentiful today and are quite popular with the "nostalgia rod" set. Shown here is a flathead with a modern 4-barrel adapted to the blower, and an Ardun-hemi Ford with a twin 2-barrel S.CO.T.

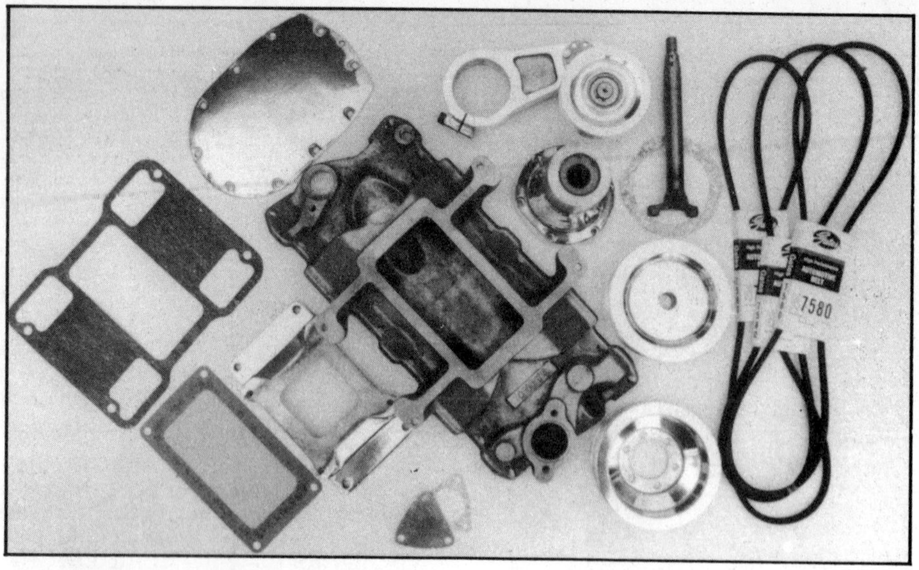

The first GMC street blower kit for small block Chevies looked like this. First made by Cragar, it was later marketed by Andy Brizio. Actually, it was quite complete; all you needed to add was a 4-71 GMC blower. This kit is still available through Cub Barnett (see GMC blower section).

Over the years all sorts of GMC-based blower cases, end plates, drives, and manifolds have been made for competition. This small block Chevy set-up with a unique 6-71 case was made by the late Pete Robinson.

THE PEPCO

Although it received little or no coverage in performance magazines, and was therefore little known to rodders, another aftermarket Roots-type supercharger produced in this country was the Pepco. The name was an acronym for Progressive Engineering Product Company (it gets a little confusing these days since Borg-Warner makes a Pesco blower, and we now also have a new aftermarket unit called a Penco). The Pepco was produced as a subdivision of Fageol Products Company beginning in the early 1950's. Lou Fageol had been very active in racing engine development for both cars and boats, and had driven in three Indy 500's in the 1940's—including a twin-Offy-powered, 4-71 GMC supercharged car! Ray Fageol, Lou's son, is now producing the K-F Roots supercharger discussed later in this book.

The reason rodders paid little attention to the Pepco was because it was designed initially for small foreign engines. The Fageols owned a Porsche dealership, were involved in sports car racing, and happened to import some of the first Volkswagens to the U.S., and Fageol employees Emmer Kelley and Art Hilf had been developing the Pepco supercharger. They decided to add one to the 36 hp VW to try to put some life into the bug, and it worked so well that they immediately discovered the inadequacy of the VW's brakes! The Pepco blowers were first displayed at Watkins Glen in 1950, installed on a Volkswagen and on an MG TC.

The Pepco was a small Roots blower with a cast aluminum case and rotors. The rotors were 3-lobed but were not twisted like a GMC. The unit was made in two sizes, a "4" and a "6," the designation referring to the length of the case and rotors in inches. From 1950 to 1958 these blowers were sold in complete kits to fit Volkswagens, MG's, and Porsches, including manifold adapters, stock carburetor adapters, and a belt drive. When VW-powered dune buggies became popular in the 1960's, many Pepco blowers found their way onto these machines.

One unusual feature of the Pepco was that it used no oil or grease. The bearings were of a special silicon self-lubricating design, and the drive gears were bi-metallic. Because they were totally oil-free, Pepcos were used to pressurize camera housings on the first U-2 spy jets, and later as air blowers in atomic welding. They were also used on some Crosley engines in outboard and inboard boat applications after Fageol acquired the engine from Crosley (later assumed by Homelite). Approximately 1000 of the Pepco blowers were made and sold, but they seem relatively scarce these days. They were a very highly regarded supercharger in their day; however, because of the special non-lubricated bearings and gears, rebuilding a worn one might be difficult today.

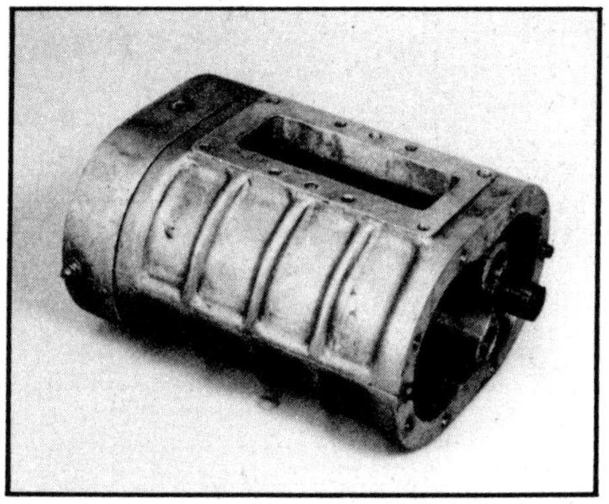

Although tons of these little Pepco Roots-type blowers were made, the only example we could find today is this partial unit with one broken rotor.

Who knows what this Roots-type compressor was originally used for, but someone obviously put it on a fuel-injected Chevy sometime in the past. Odd ball swap meet blowers like this might work fine...or they might not.

OTHER ROOTS BLOWERS

I should also point out that a wide assortment of Roots-type blowers have been produced in this country, for a variety of purposes (from scavenging or supercharging diesel engines, to pressurizing airplane or submarine cabins, to pumping molasses or flour), not to mention blowers of other designs. For instance, McCulloch built thousands of Roots-type superchargers for military landing craft powered by small engines during the war. He later sold this design to Melee-Dexter, who is still producing it. In the early 60's Dick Jones of Champion Spark Plugs used a Melee-Dexter Roots-type blower to supercharge the Offy Indy motor for an attempted, but unsuccessful, comeback. Other Roots blowers have been made by Schwitzer-Cummins, the Pesco division of Borg-Warner, and the original Roots company, just to name a few. You might very well turn up some sort of strange-looking Roots blower at a swap meet. Whether or not you could adapt it for use on an automotive engine would depend on the original intended use (primarily the rpm and pressure capacity), and whether it has sufficient clearance between case and rotors to allow high-rpm use on an automotive engine. Probably the best advice is to steer clear of unidentifiable blowers, unless the price is dirt cheap and you are capable and willing to experiment with it and modify it to suit your needs.

THE JUDSON

The only vane-type supercharger to be marketed in the United States so far is the Judson. This small compressor was manufactured by the Judson Research Company of Conshohocken, Pennsylvania, in the 1960's, and has been out of production for about fifteen years. Like the Pepco, Judson blowers were relatively unknown to hot rodders because they were designed for smaller engines, and were marketed primarily for sports cars. Installation kits were sold for Volkswagens, Volvos, Mercedes 190 SL's, Renault Dauphines, and similar cars. We will describe the Judson in more detail in a following section on vane-type superchargers; however, they are strictly a swap meet item these days.

THE LATHAM

The Latham supercharger was first introduced by the Latham Manufac-

The origin of the Roots blower on this contemporary flathead street rod is unknown, but it has been on the motor for several thousand miles. The blower is set up very loose (that is, basically worn out), so it gives little boost; but it doesn't leak oil or cause other problems, and the owner likes the looks.

turing Company of West Palm Beach, Florida, in 1956 and remained in production until 1965. About 600 units were produced during that time.

Designated an "axial-flow" blower by its designer, Norman Latham, this supercharger is actually a turbine fan similar in design and concept to a jet engine. It is the only supercharger marketed in the U.S. to use this principle for blowing air into an engine.

Basically the Latham blower consists of a stack of small fans, each made by inserting a row of blades around the circumference of an aluminum ring. These rings are bolted together to make a cylinder covered with small fan blades. This cylinder is then spun at high speed inside a cylindrical housing, to blow air from the front to the back. Such a fan arrangement, however, would tend to simply spin the air inside the housing, rather than blow it through. In order to "straighten out" the airflow through the blower, Latham placed rows of blades inside the stationary housing, turned the opposite direction from the rotor blades. Thus he created a multistage compressor, each stage consisting of a rotor fan, which accelerates the air, followed by a reverse-direction stator fan, which "catches" the accelerated air, changes its direction, and increases its pressure in a manner similar to the diffuser blades in a centrifugal blower. This compressed air is then accelerated by the next rotor fan, compressed again by the next stator fan, and so on until it reaches the back of the housing and is blown into the intake manifold at high volume and pressure. The term "axial flow" refers to the fact that the air travels along the axis of the rotor (that is, from the front of the blower to the back), rather than being spun radially by the rotating fans.

We saw this Judson blower at a swap meet recently in Oklahoma City. They are extremely rare today, mainly because most of them wore out long ago.

The Latham blower consists of a stack of stator fans, with a corresponding stack of rotor fans inside on a revolving shaft. The number of rotor and stator sections could be varied to make the blower larger or smaller. Air was drawn in through the carburetor mounts at the front (top in this photo), and emitted into the manifold of the engine at the other end. This unit has been fitted with four Carter YF sidedrafts in an unusual arrangement (two facing forward instead of all four mounted to the sides).

The Latham blower had several advantages. To begin with, it has only one moving part, the rotor assembly, and two double-row ball bearings. Other than the bearings, no parts touch, so nothing should wear out in one, plus there is very little stress on any components. The Latham design is also very efficient in terms of engine power required to drive it. Producing as much as ten pounds manifold pressure, the Latham reportedly requires as little as ten horsepower from the engine to drive it. Inventor Norman Latham claims high volumetric and adiabatic efficiency for the axial-flow design. The fact that the Latham is a draw-through configuration further adds to the cooling, plus increases fuel atomization and distribution. Latham claims that in most street installations his blower, along with Carter YF sidedraft carbs, actually increased fuel economy by as much as 12% over stock carburetion—a claim no other supercharger could make (and one which might require further substantiation).

However, there is little argument that Lathams could work well. Karol Miller used one on a '56 Ford Y-block to significantly better all class competitors at both the Daytona Beach and Bonneville speed trials—and drove the car there and back from Texas on both occasions! Les Ritchey installed one on a 312-inch '57 Ranchero, improving quarter mile times from 15.2 sec./90 mph to 13.44 sec/103 mph—and drove it from L.A. to the Oklahoma City Nationals and back. GM was very intrigued with the Latham blower in the late 50's, installing it on a couple of Motorama dream cars. A GM dyno test of a small block Chevy at that time produced 230 hp at 6000 rpm (with a single four-barrel) and netted 334 hp at 5600 rpm after the Latham was installed.

The major drawback to the Latham was its cost (about $1000 for a complete kit in the 50's and 60's) necessitated by the extremely high cost of the considerable machining and assembly processes in its construction. Even at that price, the Latham wasn't profitable to build. Another problem with the Latham was the drive belt. This blower ran at five times crank speed, which necessitated extremely high belt velocity. Latham used a flat, non-toothed, steel-impregnated belt about an inch and a half wide on a large aluminum crank pulley and a very small blower drive pulley. It was critical to get the blower pulley and crank pulley exactly aligned and parallel to keep a belt on the blower. Plus the belts wore out and had to be changed regularly.

Initially the Latham gained a bad reputation with hot rodders because they tried using it on high-revving dragster engines. The unit was designed primarily for street use, capable of delivering a stout 10 pounds boost at 4000 engine rpm, but having a practical limit of about 5000 engine rpm (25,000 blower rpm). The biggest problem at higher rpm is the belt.

Latham blowers were made in at least a dozen styles, with complete installation kits to fit most American cars of the mid to late 50's and early 60's. The major difference in blowers was the length of the "case," which was actually determined by the number of rotor and stator rings combined to make the unit. The largest Lathams were 11-stage, and these were the most common. The small block Chevy blower was a 10-stage, and smaller units were made for other engines, including the Chevy Corvair. The Lathams also came with either two or four sidedraft Carter YH carburetors. Some kits included a low profile intake manifold to fit the blower under a stock hood; others bolted onto an existing 4-barrel intake. If you can find a used Latham, chances are that it will be in good shape, and it would not be difficult to adapt it to nearly any engine.

A very interesting development, learned shortly before this book was produced, is that a company in California has just purchased rights to the Latham blower and intends to market it once again. Rather than machining the fans, they plan to cast the rotors and stators to make the unit considerably more affordable and practical to build. A new triple 5-mm V-belt drive system should solve pulley alignment and belt slippage problems. Initially a kit will be designed for new Camaros and Corvettes, using a single 4-barrel intake on one side of the blower housing.

Although some Lathams used just two carbs, this is a very typical installation with four Carter sidedrafts. Most Chevy small block applications were adapted to a Rochester fuel injection manifold, because of its low profile. As you can see, the whole package fits under the hood of an early 'Vette.

Latham blowers were certainly unusual and intriguing, and the craftsmanship in their machining and construction was superb. Here one has been installed on a Pontiac factory concept car for 1959.

CHAPTER 4

HOW BLOWERS WORK

THE IMPERFECT WORLD OF SUPERCHARGING

One of the first and most basic tenets of hot rodding is a strong distrust or downright disbelief in theory, as opposed to practical application. The modus operandi of hot rod engineering is to look at what has worked the best in the past, figure out a few practical tricks to make it work a little better, and then bolt it on the car and give it a try. If it doesn't work better, you go back to what you know through experience does work.

Although the field of automotive performance modification is becoming more and more sophisticated these days, hot rodders still believe in what they can feel when they get behind the wheel, or in what the timing slip says at the drag strip. The manufacturers of speed parts these days rely on flow benches, dynamometers, and computers to design and test new products, but the average rodder—as well as most professional dragster pilots—still assert that "you can't drive a dyno."

What these hot rodders know, and have known since the 50's, is that superchargers—especially GMC Roots-type blowers—work. You bolt one on an engine and it makes horsepower. Spin it faster, or bolt on a bigger one, and it'll make even more horsepower. Go too far and you'll blow head gaskets or melt pistons. Hot rodders have also learned that, although turbochargers are somehow supposed to be "better" than belt-driven blowers, they just don't deliver the same instant horsepower on the street or at the drag strip.

Although a dyno test may not tell everything about a blower's performance—and it doesn't necessarily reflect how the blown engine will operate in a given vehicle on the street—it is still the best way to measure how much net power a particular supercharger adds to the engine. One thing any dyno test will immediately tell you is that blowers *work*, and that belt-driven blowers are readily adjustable (by increasing or decreasing the overdrive ratio)—just turn up the boost to turn up the horsepower!

Unfortunately this quixotic practicality common to hot rodders can be both commendable and misleading. Hot rodders have done things with cars and engines that have left the theoretical engineers flabbergasted. Rodders have often succeeded where a highly-educated designer would never have attempted a project, just because the rodder didn't know that it wasn't supposed to work. If a perpetual motion machine or a perfectly efficient internal combustion engine could be built, it would be a hot rodder who would figure out how, through sheer perseverance and enthusiasm.

But the harsh reality is that a perpetual motion machine will remain an impossibility, and even the best internal combustion engine will be *far* from perfect. The laws of conservation of matter and energy are true: you cannot get more out of a system than you put into it. Second, and more important, a law called entropy applies: you can never even get *as much* out of a system as you put into it when you are converting matter to energy, and energy to work, as you are in an engine.

This is all very pertinent to supercharging because a blower can increase the volumetric efficiency of an engine above 100%. Second, although a belt-driven blower uses up a certain percentage of the engine's horsepower in order to run, it produces much more power in the engine than it takes. These facts can fool us into thinking that a supercharger is some magical apparatus which can defy the laws of physics. It can't.

The reason why a blower seems to work miracles on an internal combustion engine is because the engine is so inefficient to begin with. A good gasoline engine is lucky if it can convert 30% of the energy contained in the fuel into actual work. The addition of a supercharger, in most cases, actually *lowers* this figure (because it requires a greater percentage of fuel to make more power in the motor...we'll explain in a moment). And, as I pointed out earlier, superchargers are similar to internal combustion engines in that they are both types of air pumps. All superchargers are themselves far from 100% efficient in terms of the amount of energy they consume compared to the amount of air they pump, and in terms of the mass (weight) of air they pump compared to the size of the blower and the volume of air it displaces.

We will not be able to fully explore the intricacies of supercharger design and efficiency here. In fact, a surprisingly small amount of research has actually been compiled on superchargers, especially Roots blowers, and the handful of existing books and the several engineers I consulted during this project either had few answers or openly contradicted each other. What I want to do in this chapter is give you just enough engineering background to understand what blowers can do and what they can't, to dispel some myths about superchargers, and to give you a more solid basis from which to choose the type of blower system that will be best for your vehicle and the way you drive it.

DENSITY, VOLUME, AND PRESSURE

Myth number one is that a blower's primary function is to increase the volume or the pressure of the air in the intake system of the engine. The real purpose of the supercharger is to increase the *density* of that air.

Density is the mass, or weight, of a substance in a given volume. For most solids and liquids, weight is nearly proportional to volume—increase the volume, and you increase the weight a proportional amount. In other words, the density of a given solid or liquid does not change very much under normal conditions. Consequently we tend to think of an increase in volume of a substance as an increase in the amount (mass) of that substance.

But this thinking is never wholly correct, and it certainly does not apply to gases. The density of a gas, such as air, changes considerably as its temperature or pressure changes. Increase the temperature, and the density of the gas decreases (if it's not in a closed container), because the gas expands. Increase the pressure, and the density will usually increase, because a greater amount (mass) of the gas will be compressed into a smaller space (volume). I say *usually* because we are dealing with more than one variable at the same time when we are dealing with the state of a gas. If a gas is both compressed (which should increase its density) and heated (which should decrease its density) concurrently, the net result could be an increase, a decrease, or no change at all in the density of the gas.

These facts are extremely important to an understanding of supercharging because a supercharger does in fact both compress and heat the air entering the engine, in addition to increasing the volume of air passing through the intake system.

THE GAS LAW

Before we go any further, let's take a quick look at the basic law which describes the relationship between the pressure, volume, weight, and temperature of a gas. Known as the Ideal Gas Equation, or just the Gas Law, it is usually stated by the equation:

$$PV = nRT$$

where P is pressure, V is volume, n is the weight of the gas in moles, R is a constant (called the Molar Gas Constant, which has different values depending on the units used for the other quantities), and T is the temperature. It

What size of blower do you use on what size of engine? The question has more ramifications than are immediately apparent, but for positive displacement blowers, the basic answer is logical—match blower displacement to engine displacement. Thus a big 6-71 GMC will feed the Arias Chevy, a medium-sized 3-71 will handle a V-6, and a pint-size S.CO.T. will give the tiny Crosley dragster engine all the pressure it can handle.

is not within the scope of this book to fully explore how this equation can be used, nor how gases act in general—you can find an explanation of the gas law, how to work it, and values for *R*, in any thermodynamics textbook. What this equation should immediately show you, assuming you have a basic familiarity with algebra, is the relationship between the pressure, volume, temperature, and weight of a given gas.

If we are discussing a given amount of gas, in moles (that is, a certain weight of the gas...or a certain number of molecules of the gas, if you want to look at it that way), then both *n* and *R* in the Gas Law would be constant, and we could write the equation thus:

$$\frac{PV}{T} = \text{CONSTANT}$$

What this tells us is that for a given amount (weight) of gas, increasing the pressure and keeping the volume the same will increase the temperature proportionally; increasing the temperature and keeping the volume the same will increase the pressure proportionally; increasing the volume and keeping the pressure the same increases the temperature, and so on.

In relation to supercharging, the significant things the Gas Law tells us is that increasing the pressure increases the temperature, and vice versa; that a temperature rise at a given pressure will increase the volume, which means a decrease in density; and that both the pressure and the temperature could be increased without increasing the density.

THE PROBLEM OF HEAT

The thorn of supercharging is heat. Heat is a very common and easily manifested form of energy—a very easily wasted form of energy—which is a typical byproduct of work being done. In supercharging, heat is a multiple detriment. In the first place, the act of compressing air heats it, as we saw above. Second, much of the engine horsepower used to turn the supercharger and compress the air eventually is converted to more heat which is transferred to the air. Most blowers, especially Roots types, literally beat the air, which heats it considerably; plus friction between the air and stationary and moving surfaces, as well as the friction of air turbulence, all combine to heat the air further. The heat tends to expand the air (increase its volume), which raises the pressure,

The beauty of belt-driven blowers is that they are adjustable. Just change the pulley ratio to dial in the boost you want. Thus, if you want a big, impressive, polished 6-71 on your small block Chevy you can underdrive it slightly to give practical street performance, as blower builder Don Hampton has done on his wife's street T. Such a set-up may not give maximum efficiency, but losses are minimal compared to overall power gains.

but does not increase the density.

To give you a practical example of what could be taking place in a supercharger, imagine a closed box full of air at room temperature and atmospheric pressure. Let's say the box measures one foot per side, so that its volume is one cubic foot. The volume of air in the box, one cubic foot, cannot change, since no air can get in or out. Likewise the density of that air—the amount of weight of air, or the number of molecules of air, per cubic foot—cannot change. If we insert a pressure gauge through one of the walls of the box, it will read zero psi, since both the pressure and the temperature inside the box is the same as that outside (most pressure gauges read in pounds per square inch above atmospheric; thus a 5 psi gauge reading is really approximately 19.7 psi absolute, since atmospheric pressure is about 14.7 psi at sea level).

Now let's take our box of air and put it on the stove. As the air inside is heated it tries to expand, but cannot because it is contained by the box. The result? The pressure increases. We will actually see "boost" on the pressure gauge, even though the air is not being compressed, and *the density remains exactly the same.* In other words, we see a boost reading on the gauge even though the amount of air in the box has not increased at all. The same thing can happen, in extreme cases, in the manifold of a supercharged engine. If a blower heats the air drastically, instead of pumping it efficiently, it could show a boost reading on the gauge—simply by *expanding* the air, without causing any increase in the volumetric efficiency—or power—of the engine.

To see how this could happen, we can use the Gas Law in its simplified form:

$$\frac{P1\ V1}{T1} = \frac{P2\ V2}{T2} = \text{CONSTANT}$$

Knowing that the volume is constant in both cases, we can figure the pressure rise by using the proportion:

$$\frac{P1}{T1} = \frac{P2}{T2}$$

in which temperature must be figured in degrees Rankine (absolute temperature, which equals degrees Fahrenheit plus 460), and pressure in psi absolute. Using this simple equation, we find that if our box of air starts out at typical atmospheric temperature and pressure of 60° F and 0 psi (520° R and 14.7 psia) and is then heated to 240° F, we will see a "boost" reading on the pressure gauge of 5.1 psi. Remember, we have neither compressed the air nor increased its density—we have simply heated it—and we have five

If the carburetor inlet on your blower draws air from inside a closed engine compartment, as is the case on this Forty Ford, it can be drawing excessively heated—that it, less dense—air into the blower, which counteracts the supercharger's purpose (to make the air denser). You may also notice that this street rod uses a firewall-mounted water injector to counteract detonation in the engine—which is a byproduct of a heated intake charge.

pounds of boost on our gauge!

Now, although a blower manifold is not a closed box, nor is real air an ideal gas, the above example still gives a good approximation of what *could* happen in a blown motor. If you were getting a 200° F temperature rise in the manifold at 5 pounds of boost, your supercharger would be giving you no increase in charge density at all, and would be robbing your engine of horsepower at the same time that it would be increasing intake temperature, which would lead to detonation. Most superchargers work considerably better than this example, but all sacrifice a good percentage of their potential benefit to the demon of heat.

BLOWER EFFICIENCY

Since machines such as engines and superchargers do not operate perfectly in the real world, we use the term "efficiency" to compare how well they *do* work to how well they theoretically *could* work in a world where perfection were possible. Blower manufacturers and blower salesmen often refer to a single "blower efficiency" figure, but this is very misleading. Like the internal combustion engine, which can be measured for volumetric, thermal, mechanical, and other efficiency ratings, the supercharger likewise must be examined for efficiency in terms of different functions. Although the adiabatic, volumetric, and mechanical efficiency ratings of a supercharger all do influence one another to some degree, it is not really practical to try to lump them all together into one comprehensive blower efficiency figure. One big problem is that the mechanical and volumetric efficiencies of a supercharger operating on a real engine are very difficult to measure practically or accurately. Consequently, and because it is the figure that has the greatest bearing on supercharger performance, the efficiency figure given for blowers is usually the adiabatic efficiency, which can be readily calculated. However, if someone gives you a blower efficiency figure for a given supercharger, ask what sort of efficiency was measured and how the figure was determined.

ADIABATIC EFFICIENCY

If you are not familiar with thermodynamics, the term adiabatic efficiency may sound scary or complicated, but it's really quite simple. The problem is that many blower "experts" have used the term without fully understanding its meaning or its computation, and have therefore spread some confusion about adiabatic blower efficiency. To the non-theoretical blower builder or user, the adiabatic efficiency of a blower is a measure of how much it heats the air that it is pumping. The less it heats the air, the more efficient it is. As we have seen above, and as we will discuss further in a moment, the amount a blower heats the air charge has a large (negative) effect on the amount of power it will produce in the engine, plus it is a good indicator of how well the blower is doing the job of pumping (that is, it reflects mechanical and volumetric efficiency losses) since extraneous work done on the air by the blower will be converted to heat.

Since the blower is going to heat the intake charge as it compresses it, no matter what you do, one of the easiest ways to keep intake charge temperature down is to draw cool air *into* the blower by means of a fresh air scoop or duct. In other words, regardless of the adiabatic efficiency of your blower, an easy way to lower T2 is to lower T1.

The temperature rise across the blower (the difference between inlet air temperature and outlet, or manifold, air temperature) is easy to measure. Consequently it is easy to compare the temperature rise with one blower at a given boost pressure on a given engine to that of a different blower at the same boost on the same engine. Such a comparison will certainly give us an approximation of the relative adiabatic efficiencies of the two blowers, but it won't give us a correct percentage figure.

One recent book on superchargers states that a blower which draws in air at 70°, compresses it to 1.5 atmospheres (about 7 pounds of boost), and discharges it at a temperature of 70° would have 100% adiabatic efficiency. As we saw from the Gas Law, this is impossible unless we incorporate some very effective cooling device into the supercharger. The act of compressing air heats it (because we must do work on that air to compress it, and the air "absorbs" the energy of that work as heat).

If we could compress a gas by some physical means without adding any more heat to it than the heat generated by the act of compression, and we did not let any of that heat escape, we would have perfect adiabatic compression, or 100% adiabatic efficiency. (The term "adiabatic" simply means a thermodynamic process in which no heat is gained from, or lost to, sources outside the system.) Although perfect adiabatic compression is impossible in the real world, we can calculate the heat of perfect adiabatic compression by the following formula:

$$T2 = T1 \left[\frac{P2}{P1}\right]^{.283}$$

In this formula *T1* is the inlet (or ambient) air temperature, *P1* is the ambient air pressure, *P2* is the outlet (or manifold) pressure, and *T2* is the new temperature to which the act of compression will raise that air. Temperature is figured in degrees Rankine and pressures are absolute. To get *T2* in degrees Fahrenheit, subtract 460.

Raising a number to the .283 power is a bit difficult, but the pressure ratio *(P2/P1)* raised to this power can be quickly figured from "Y-tables" for air in some thermodynamics textbooks or handbooks. Once you have figured the ratio of (P2/P1), *the Y-table will give you a value of Y for this amount (for instance, if P2/P1* is 1.5, or about 7.4 pounds boost, the value of *Y* is 0.12159). Then using the simpler formula:

$$\Delta T(ideal) = T1 \times Y$$

you can figure the ideal temperature increase, in degrees Fahrenheit, for a compressor at 100% adiabatic efficiency.

To show how the equation works, let's say that the ambient air temperature is 68° F; this is the blower inlet temperature, or *T1*, which converts to 528° R. Ambient air pressure, *P1*, is one atmosphere, or 14.7 psi absolute. Let's say that the blower is making 7.4 pounds boost in the manifold, so *P2* is 22.1 psi absolute. If our blower were a perfect compressor, the formula shows us that it would heat the air 63.36° F, or raise it to a *T2* of 131.36° F. If we raise blower boost to 14.7 psi (29.4 psia), *T2* would go up to 182.42° F. This would be the temperature rise in the manifold if the blower had 100%

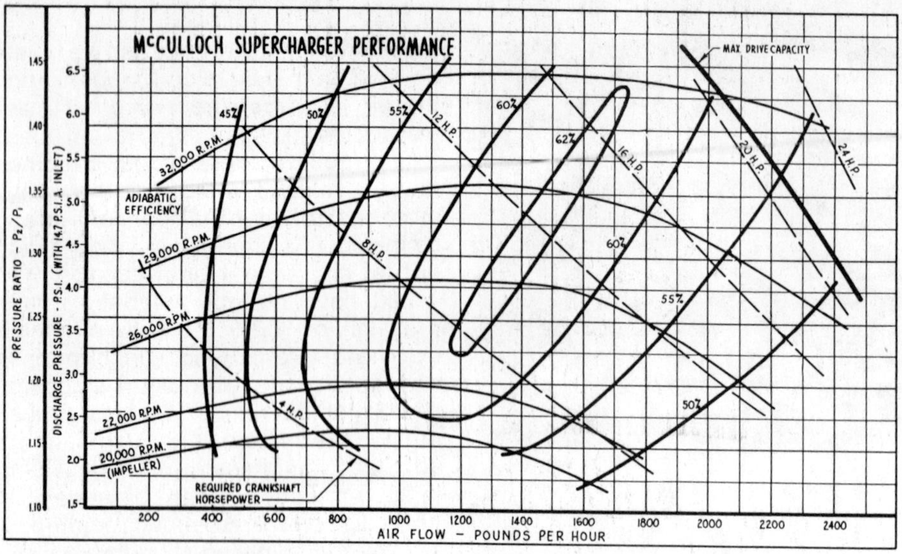

A compressor "flow map" for a Paxton/McCulloch centrifugal supercharger shows adiabatic efficiency and horsepower required to drive the blower (mechanical efficiency). The ring near the center of the map—known as an "island"—indicates the range of this blower's maximum adiabatic efficiency, in this case 62%.

adiabatic efficiency. But, of course, no supercharger can have 100% adiabatic efficiency because several other factors in a real mechanical pump contribute to heating the air: friction of air against blower parts, friction of air against air, friction of mechanical parts, air leakage, and so on.

To figure the adiabatic efficiency of a given supercharger, use the equation:

$$\text{ADIABATIC EFFICIENCY} = \frac{(T2 - T1)\ \text{ideal}}{(T2 - T1)\ \text{actual}}$$

which is the ratio of the perfect (adiabatic) temperature increase figured from the previous equation, to the actual temperature increase observed across the supercharger. For instance, if we use the example above of the blower making 7.4 pounds boost with an inlet air temperature of 68° F, the ideal heat rise *(T2 - T1)ideal*, or the heat rise at 100% adiabatic efficiency, should be approximately 63° F. But let's say that the actual temperature in the manifold is 194° F. This would give us a *(T2 - T1)actual* of (194 - 68) or 126 degrees. Our adiabatic efficiency ratio would therefore be 63/126 or .50; expressed in percentage, the adiabatic efficiency of this supercharger would be 50%. Given the same inlet temperature and manifold pressure, if the actual manifold temperature were 152° F, the blower's adiabatic efficiency would be 75%.

Hardly any superchargers have an adiabatic efficiency of 75%. Most Roots-type blowers are doing well if they reach 50% adiabatic efficiency.

MECHANICAL EFFICIENCY

In this book we are dealing specifically with superchargers which are driven mechanically by the engine, as opposed to turbochargers which are driven by exhaust heat (which is engine energy normally wasted), or even to superchargers driven by some other remote energy source (such as an electric motor—one of these was actually advertised in magazines in the 1950's!). Consequently all of the blowers we will be discussing consume a certain percentage of the engine's power in order to make more power. If superchargers didn't make a lot more power than they consumed, we wouldn't use them; but you might be surprised to find out how much horsepower it takes to turn a 6-71 GMC making 15 to 20 pounds of boost.

The mechanical efficiency of a supercharger is a measure of the amount of engine horsepower the blower actually consumes to produce a certain amount of boost at a given speed, compared to the amount of power it would take to compress air at the same rate under ideal conditions.

From readily available thermodynamics equations we could calculate the amount of horsepower it would take to turn a supercharger of a given displacement, at a certain rpm, to make a certain amount of boost under ideal conditions—that is, without any mechanical losses due to friction, turbulence, leakage, or other real-world conditions. The ratio of this figure to the actual horsepower a real supercharger consumes to produce the same amount of boost would give us the mechanical efficiency percentage of the supercharger.

However, a practical and reliable method for measuring the actual horsepower drag of a blower on a real running engine would be very difficult to devise. Since most of the mechanical losses in the blower show up in the form of heat, we could figure an approximation of the mechanical inefficiency of a blower by converting the measurable extra heat energy gain (above 100% adiabatic efficiency) into work energy units (horsepower). However, this would still be a rough approximation involving too many variables—and it shows that the adiabatic efficiency figure, which is easier to calculate, is more meaningful; doubly so, since it partially reflects mechanical efficiency as well. So we will not delve into formulae here for figuring mechanical blower efficiency.

However, to give you some idea of the amount of horsepower it can take to drive a big GMC blower at high boost, we offer the following example. Bob Brownell, a California drag boat racer who runs a very successful Arias-powered blown gas flatbottom, decided to build a "blower dyno" to make some real tests of GMC superchargers. The unit consisted of a stand for the blower with a closed box for a manifold below it. In the box were three 2-

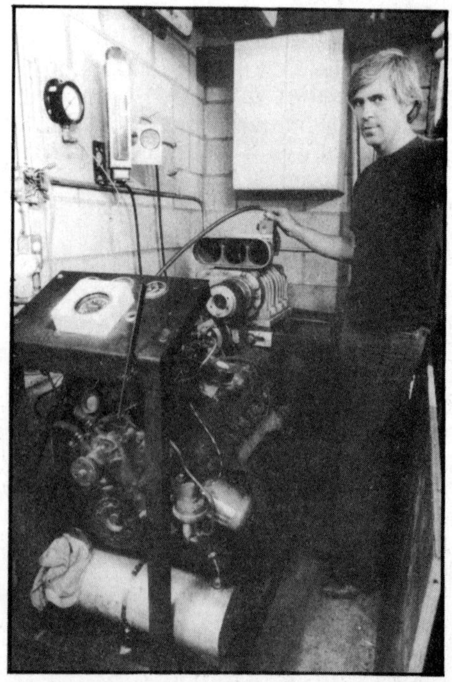

It can take quite a bit more power than you might think to turn a large supercharger, especially one that is set up with Teflon strips and tight clearances for racing. Larry Bowers built this test stand to "run in" his blowers and test them, and it uses a 351-inch Ford V-8 just to turn the bare blower.

inch ball valves to regulate the amount of air escaping from it, which would regulate the amount of pressure, or "boost," in the "manifold." This entire unit was mounted so that it could rotate on an axis about the blower drive pulley, and a torque arm was fixed to the blower case so that it could record the torque being applied to the blower on a scale. This would give a direct reading in foot-pounds of the amount of torque actually being applied to the blower to drive it, which could be converted to horsepower in the usual manner.

With an 8-71 GMC blower on this dyno, Bob first hooked up a 365-inch stroked small block Chevy V-8, with a single 4-barrel, headers, and so on, to turn the blower. But this engine would not pull it. When they closed one or two of the valves to make boost, the blower literally stopped the motor! Next Bob tried a 427 big block Chevy, and this engine could only hold 5-6 pounds boost with the 8-71 race blower before bogging down. So finally he built a 454 Chevy just to pull the supercharger and got readings up to 17 pounds of boost. Unfortunately the pulsing of the blower made torque readings on the dyno scale nearly impossible, but it was fluctuating between 200 and 400 horsepower—just to drive the blower.

Of course this is an extreme example, but it does illustrate several things. First of all, it shows how difficult it would be to really test how much horsepower it takes to drive a blower—and this set-up is still far from the situation of a blower operating on a real engine. Secondly, it does give a graphic indication of the large amount of power it could take to drive a big, tight-clearanced racing blower; naturally a smaller street blower would take far less horsepower, but it might be more than you think.

Finally, and this might be the most significant of all, Bob also recorded temperature increases with different blowers under different conditions. A chart of two tests is shown here, done with a 6-71 GMC and taken at an ambient temperature of about 75 degrees. The first test was done at a higher restriction, so that the blower made more boost at lower rpms; the second test was made with all the valves open on the manifold box, and we see that the blower heated the air more per pound of boost at the higher rpm speeds—that is, its adiabatic efficiency was lower. These tests were run with dry air (other than a light oil mist for lubrication); temperatures would be lower with fuel. However, on actual dynamometer tests of one of Bob's 512-inch blown gasoline Chevy big block engines, he recorded 340 degrees of manifold temperature at 18 pounds of boost while making 800 hp. Using an intercooler of his own design below the blower, he reduced manifold temperature to 240° and increased engine power to 900 hp. Does this tell you something?

TEST OF 6-71 GMC BLOWER

RPM	PSI (manifold)	TEMP (manifold)
TEST #1 (TWO VALVES OPEN)		
2000	1	92° F
4000	5	98° F
6000	12	200° F
7200	15	230° F
TEST #2 (THREE VALVES OPEN)		
2000	.5	80° F
4000	2	82° F
6000	3.5	100° F
8000	6.5	210° F

VOLUMETRIC EFFICIENCY

A third measurement of blower performance is volumetric efficiency. Blower volumetric efficiency is very similar to engine volumetric efficiency (remember, both engines and blowers are air pumps), and it is actually

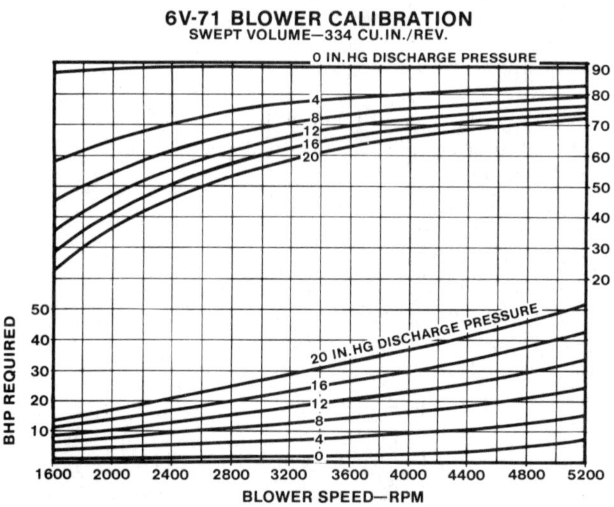

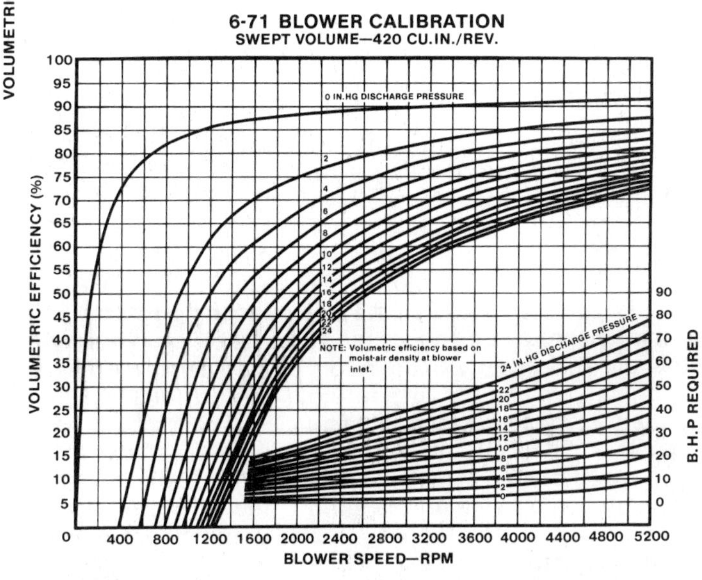

These two charts show the volumetric efficiency and mechanical efficiency ("B.H.P. Required") for both 6-71 and V6-71 GMC blowers as tested by Detroit Diesel engineers. These blowers were set to factory clearances, as shown on the 6-71 chart, and the tests were made at blower speeds from 0 to 5200 rpm. The successive curves show results at increasing boost pressures, from 0 to 24 inches of mercury. (30 inches of mercury is roughly one atmosphere, or 14.7 psi. 20 inches of mercury would be roughly 9.8 pounds of boost at the blower outlet.)

measured the same way. Specifically, the volumetric efficiency of a supercharger is the ratio of the actual volume of air it pumps per revolution, compared to its internal displacement per revolution.

I feel that supercharger volumetric efficiency is perhaps the least understood, or just the most difficult to pin down, of blower ratings. In the first place, the internal capacity of a positive displacement blower, such as a Roots, is easy to calculate, since there is a measurable swept volume per revolution (in a Roots, it's the volume contained between two rotor lobes and the case, times the number of lobes which "circulate" in one revolution). For instance, the displacement of a 4-71 GMC is 274 cubic inches and that of the 6-71 is 411 c.i. However, the static internal volume of non-positive-displacement blowers such as centrifugal or axial flow types is nearly impossible to figure, especially for the purpose of deriving a volumetric efficiency number.

Here we have examples of two nearly identical blower installations—in this case, Dyer V6-71's mounted on small block Chevies with dual AFB 4-barrel carbs. The only real difference is that one has a large air scoop on the carbs, while the other has two small air cleaners. Now, you certainly should run air cleaners on any street blower, but if that big blower really starts to draw some air through the carbs, chances are that these little paper-element cleaners are not going to be able to keep up with the flow. The restriction, or drag, they cause on the blower inlet will greatly reduce the blower volumetric efficiency.

For sake of measurability, supercharger testers figure blower *VE* as the ratio of air volume drawn into the blower to blower displacement, rather than trying to measure the actual pumping volume on the outlet side. The volume of air being drawn into the blower inlet can be measured, in cubic feet per minute, with a manometer-type airflow gauge, as used to measure airflow through carburetors on test engines.

Whether we measured the air going into the blower, or the air coming out, either way we could get figures to compare the pumping efficiency of a given blower under varying conditions, or of different blowers under the same conditions. But trying to determine a numerical volumetric efficiency for any one supercharger, especially compared to other types of superchargers, is difficult and usually misleading.

The problem is, once again, that we have several variables operating at the same time. Most textbooks, and some engineers involved with superchargers, contend that the volumetric efficiency of a Roots-type blower increases with its rotational speed. They base this assertion on the observation that air leakage around the rotors is the primary volumetric loss in the supercharger. Since the amount of air that can leak past clearances around the rotors is a function of time (it takes a certain amount of time for the air to leak through the clearances), the faster the rotors turn, the less leakage can occur per revolution; consequently, the higher the blower rpm, the higher the volumetric efficiency.

On the other hand, with any type of supercharger, but especially with a Roots or other backflow compressor, as manifold pressure increases with blower speed, volumetric efficiency will decrease because of backpressure, which causes increased leakage.

Also, remembering that superchargers are air pumps similar to internal combustion engines, it stands to reason that blowers would suffer the same volumetric deficiencies that engines do. That is, any blower has a given size intake "port" and what we might call "cylinders," or air pumping cavities. Although we have the supercharger to force feed air to the engine, we still have only atmospheric pressure to feed air into the blower. As the rotors in a Roots blower, for instance, turn past the inlet "port" opening, there is only a given amount of time for the atmosphere to push air into the rotor cavity before it "closes" (turns past the opening to seal against the case). The faster the rotors turn, the less time they will have to fill completely with air, and the lower will be the blower's volumetric efficiency.

The problem is to determine accurately, *under actual operating conditions*, which of the several factors affecting blower volumetric efficiency is predominant, or how they combine for a net *VE* under various conditions. This might be more important than it at first seems. At least one new Roots-type blower has been designed primarily around the assumption that volumetric efficiency is the major design criterion, and that blower *VE* is predominantly a function of air leakage around rotor clearances—at least in street boost ranges up to 10 psi. Given this assumption, the conclusion is that as small a blower as possible should be used on an engine for two reasons: (1) because you want the blower to run as fast as possible to make a given boost (in this case, it's designed to be driven nearly 100% faster than the engine), and (2) since air leakage is not only a function of time, but also of total clearance surface area. Making the blower smaller, or just making the rotors shorter, will reduce the size of the air leakage path and therefore improve volumetric efficiency.

There is certainly logic behind this thinking; however I think it fails to take into consideration enough of the variables determining overall blower efficiency to make it the primary factor in sizing a blower and determining its overdrive ratio. There are logical reasons for stating that a larger blower, turned at a slower speed, with tighter rotor clearances, would be the better choice. The slower blower rpm would allow setting the rotor clearances closer, since there would be less rotor growth and, in GMC's, less rotor "untwist." If the blower turned slower relative to engine speed to make the same

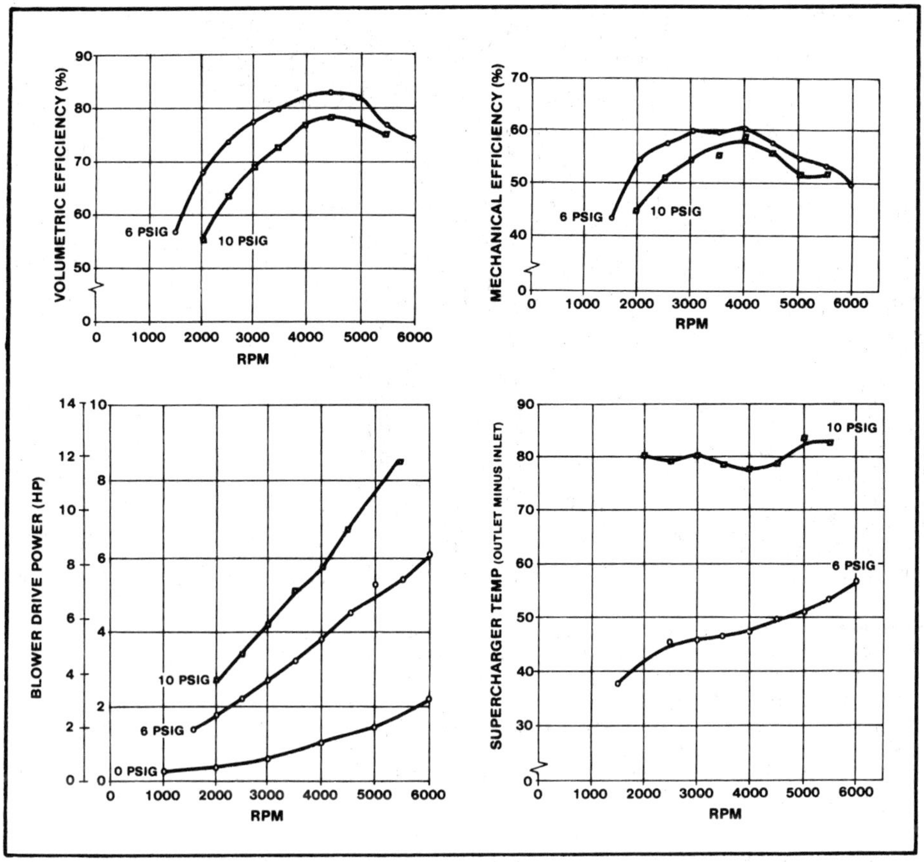

These four graphs plot volumetric efficiency, mechanical efficiency, power required to drive the blower, and temperature rise between blower inlet and outlet for a Bendix 70 cubic inch vane-type supercharger. We do not know exactly how the tests were conducted or how the measurements were taken, but somehow blower outlet pressure was maintained at constant readings of both 6 psi and 10 psi throughout a wide rpm range (in actual use on an engine, most blowers will increase boost with increasing rpm). However, the general shape of these curves is indicative of typical supercharger performance. And it is particularly interesting to note the differences between a blower operating at 6 pounds boost and one at ten pounds.

boost, it would be beating the air less, or doing less extraneous work on it, which would heat the air less (that is, raise the adiabatic efficiency). This is not only desirable in itself, but it would also heat the blower parts less, reducing expansion, and allowing tighter clearances for better *VE*. Further, although tighter clearances will slightly increase the power needed to turn the blower, a longer blower turned at slower speed (such as a 6-71 rather than a 4-71) will produce more boost with the same, or less, drag from gears, pulleys, belt, and so on.

The purpose of the above illustration is twofold. First, it demonstrates that the science of blower technology requires a close examination and evaluation of many variables before accurate assumptions can be made. Worse, because so many variables are involved, it is often difficult to determine exactly what change—or what side-effect of a change—led to an observed increase or decrease in blower performance. Second, it indicates that a good formula or rule of thumb for choosing the proper size supercharger for a given engine will be impossible to give.

I was hoping that I would be able to discover or devise some sort of table or equation which would immediately tell you how large the optimum blower should be for a given displacement engine. But once again, a broad array of variables has thwarted us. In a practical sense, the sizing of a supercharger to an engine is not as critical as one might think—certainly not as critical as the sizing of a turbocharger. For one thing, the mechanical drive of the blower can be stepped up or down easily to help regulate the output to the engine's needs. Manufacturers of the smaller blowers, such as the single Magnuson and the KF, can tell you from experience which size of their blowers will give what boost levels on various small-displacement engines. More critical in proper sizing are the centrifugal superchargers, but here the problem is quickly solved, since the only two centrifugal blowers currently made come in only one size each.

The GMC blowers are the ones that cause the most concern, however. Should you use a 4-71 or a 6-71, or possibly some other size? As Gene Mooneyham put it, "The only thing that really makes any difference is how big of a blower the customer wants sitting on top of his engine." What Gene means is that he, or any other really good GMC builder, can set up either size of Jimmy to work properly on large or small V-8's. It's a matter of setting the clearances and using Teflon seals or not. But we'll cover this in the next chapter.

THE AIR DENSITY RATIO

The biggest problem with using blower volumetric efficiency as a guide in designing or sizing a supercharger is the same problem we described earlier about confusing air *volume* with air *density*. Even if we could accurately measure the volume of air being pumped out of the blower, this is not really an accurate measure of the increase in air the blower is supplying to the motor. Why? Because, as we saw earlier, air is elastic—differences in temperature or pressure will change the volume of a given amount of air; and this is exactly what happens on the outlet side of a supercharger.

It is not an increase in volume, but an increase in *density* of intake air that increases power in an engine. We want to pack more molecules of air into the cylinders so that (1) we will have more molecules of oxygen to combine (burn) with more molecules of fuel, and (2) we will have more molecules of matter per volume in the cylinder to expand and exert pressure on the piston after combustion. We measure the number of molecules of a gas in a given volume by the weight of the gas per volume, and this is called the density.

As we saw from the Gas Law, the density of a gas depends on both its pressure and its temperature. Since these are quantities we can readily measure for the intake charge, with temperature and pressure gauges, we can figure the actual air density increase being produced by the supercharger in the intake manifold. Although the air density ratio is a term not often mentioned in relation to superchargers, I feel it is a much more meaningful (not to mention more easily calculated) measurement than the blower's adiabatic or volumetric efficiency. The formula for calculating the air density ratio is shown at right.

$$\text{DENSITY RATIO} = \frac{\text{INLET TEMP ABSOLUTE}}{\text{OUTLET TEMP ABSOLUTE}} \times \frac{\text{OUTLET PRESSURE ABSOLUTE}}{\text{INLET PRESSURE ABSOLUTE}}$$

If we use the previous example of a supercharged engine operating at an ambient temperature of 68° F and standard atmospheric pressure (14.7 psi absolute), making 7.4 pounds of boost in the manifold and recording a temperature in the manifold of 131° F, we can plug in these figures and find the density ratio:

$$\text{DENSITY RATIO} = \frac{68 + 460}{131 + 460} \times \frac{7.4 + 14.7}{14.7}$$

$$\text{DENSITY RATIO} = \frac{528}{591} \times \frac{22.1}{14.7}$$

$$\text{DENSITY RATIO} = .8934 \times 1.5$$

$$\text{DENSITY RATIO} = 1.34$$

The air density ratio measured across the supercharger for this engine is 1.34 to 1, or a 34% increase. The beauty of this equation is that it takes into consideration most of the real factors operating on the intake charge, such as the cooling effect of evaporating fuel in the air.

You can figure the air density ratio in your engine easily by installing a manifold boost pressure gauge and a manifold air temperature gauge in your car. No supercharged vehicle should be without these two instruments. To get accurate calculations, you should also know the ambient air temperature and pressure, for which you would need a thermometer and a barometer (if the blower intake is not routed to fresh air, you need to know the actual underhood temperature—but you *should* be drawing in fresh air to the blower). For rough calculations you could certainly approximate outdoor temperature and figure ambient pressure at 14.5-14.7 psia.

Since engine power is roughly proportional to intake air density, the blower air density ratio will tell you how much increase in horsepower the supercharger is providing, minus the horsepower consumed by the blower to operate. For instance, if the engine above made 200 hp unblown, the blower would increase horsepower 34% to 268 hp, minus let's say 15 hp to run the blower, for a net horsepower of 253, or a net power gain of 26.5%. These figures were not taken from any real engine, remember; they are simply examples to show how the calculations work. On the other hand, they are not all that far off from typical street blown engines. The point is, although anybody can tell you a blower will double or triple your horsepower, you don't know what it is really doing unless you put the engine on a dyno—or much more simply, install boost and temp gauges and figure the above calculations.

THE LIMITS OF SUPERCHARGING

What the blower density ratio tells us is that you can pump more horsepower into your engine in direct proportion to the increased air/fuel charge you pump into it. That's the undeniable beauty of supercharging—it overcomes the natural limits imposed by a normally aspirated engine's volumetric efficiency. But there are obviously limits to the amount of power we can pump into a motor with a blower. If there weren't, we would use small, compact engines in all of our cars and supercharge them to make 200 or 300 horsepower.

There are three basic limits to supercharging. First is the point of diminishing returns of the blower itself, which is determined by its overall efficiency. At some point the blower will not be able to make any more boost because of pumping losses, or it will heat the air more than it compresses it, or it might even reach the point of consuming more engine horsepower than it is providing. Some of these problems become factors in blown competition engines, but they rarely affect street supercharging where boost levels will normally be less than 10 pounds. The only exception would be trying to supercharge a large displacement engine with much too small a blower.

One of the few manufacturers of Roots-type blowers to try to improve the volumetric efficiency of this type of supercharger is Jerry Magnuson. This is a very early version of one of his Magnachargers, but note that it incorporates a special airflow divider bar at the inlet (arrow). This, along with a wider opening, helps to route airflow directly into rotor cavities while keeping the upward-moving portions of rotors, in the middle of the blower, from upsetting the incoming flow.

Second, the amount of boost you can pump into the engine, and the resulting increase in cylinder pressure, will be limited by the physical strengths of engine components. It does no good to raise boost beyond the point where you blow gaskets, burn pistons, or break cylinder walls. In a competition engine, you can increase the strengths of these components to a point to withstand greater pressures or temperatures. But, again,

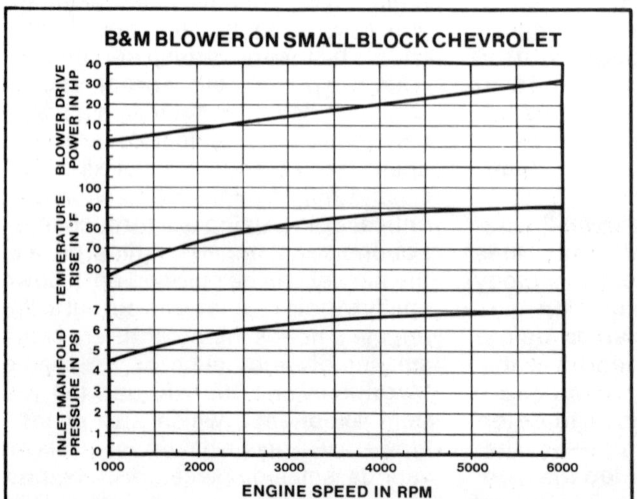

This composite graph shows the projected performance of a 144 cubic inch B&M supercharger on a 350-inch Chevy V-8. The bottom line shows the blower making 5 to 7 pounds of boost, increasing with engine speed. The middle curve shows heat rise across the blower, which roughly parallels the boost curve. The top line shows horsepower required to drive the blower, which appears to increase linearly with speed.

most street supercharging applications do not reach the physical limits of even stock engine components.

For street supercharging the practical limit to horsepower gain is detonation. Once an engine begins to detonate (knock), its horsepower curve stops climbing. And not only does the detonation point of an engine limit power output beyond that point, but serious detonation in the cylinders will also quickly destroy engine components such as pistons, rings, and spark plugs.

Two primary factors determine the point at which an engine will detonate: the effective compression ratio (or peak cylinder pressure) and the octane rating of fuel used. In a street supercharged engine, the fuel will be pump gasoline, which comes in dismal octanes these days. (The octane number of a fuel is simply and literally a measure of the point at which it begins to detonate in a special test engine with a variable compression ratio—the higher the number, the higher the compression before it knocks.) Since part of the reason for adding a supercharger to a motor is to increase the effective compression ratio, and since the blower will also raise the temperature of the charge in the cylinder (which also abets detonation), other means will have to be taken to control detonation if the supercharger can be used to raise compression to a level where it will significantly increase power.

First of all, naturally, you will want to use the highest octane gasoline available. Better yet, you can mix a 50-50 blend in your tank of unleaded premium and a leaded regular to get a blend with a higher octane rating than either. Further, you could use gasoline additive "octane boosters" available in cans at speed shops, but this is both bothersome and expensive on a regular basis if you drive your car a lot.

Other tricks which help control detonation in street engines include richening the fuel mixture (to cool the intake charge), retarding the ignition, or injecting water into the intake airstream. Each of these methods is practical and effective on a blown streeter, but you should realize that each is a compromise. With any of the above "bandaids" you will be getting more power than you could if the engine goes into detonation—but not as much as you would if the engine did not otherwise detonate at that point (for instance if the fuel octane rating were higher, or the intake air temperature were cooler). We will discuss ignition timing, carburetion jetting, water injection, and nitrous oxide injection for blown motors in following chapters.

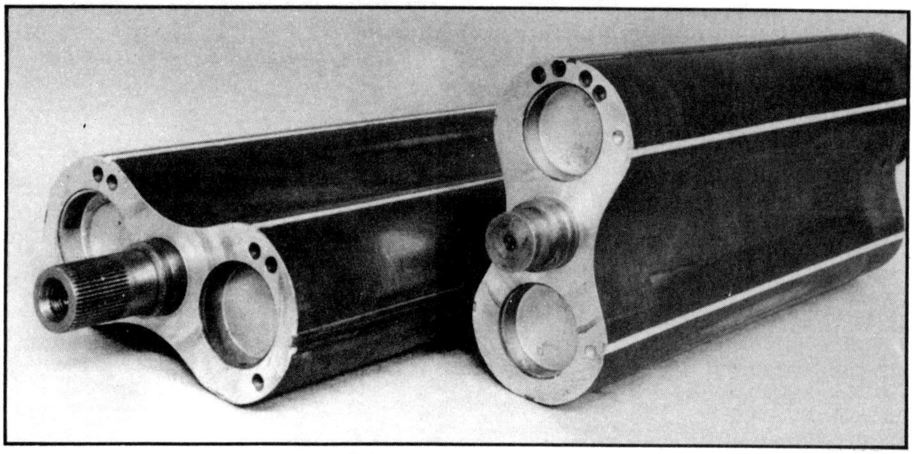

Lately the most practical and efficient means for increasing volumetric efficiency in Roots blowers has been to install Teflon or Nylon strips along the sealing edges of the rotors. However, this also reduces the mechanical efficiency of the blower (by making it harder to turn), plus it is not very practical on street blowers since the strips wear out and have to be replaced. Also, as these examples show, various GMC blower rebuilders have been experimenting with 2-lobe, straight-cut rotor designs for years to improve the pumping capacity and air/fuel distribution of Jimmy blowers.

The point is that on a street blown motor, running on gasoline (as opposed to race motors running on much "colder" fuels like alcohol or nitromethane), the primary factor limiting the amount of boost you can pump in with the supercharger will be the octane rating of available gas, rather than the melting or breaking points of engine parts.

As a matter of fact, one seldom-mentioned benefit of supercharging pointed out by British author L.J.K. Setright in his book on Turbocharging and Supercharging is that the mean effective pressure in the cylinder of a supercharged engine is considerably higher than that of an unblown motor (it has to be for the engine to make more power, since engine power is a function of MEP), but the peak cylinder pressure allowable in either engine cannot exceed the physical strength limits of the engine parts (or the knock limit of the fuel). This first of all demonstrates why a supercharged engine can make more power than a naturally-aspirated engine of the same effective compression ratio (that is, why it is better to lower the static compression ratio and add a blower rather than to use high compression pistons to achieve the same effective compression in the cylinder). The answer is that the blower produces more pressure in the cylinder *throughout* the power stroke, whereas the high compression piston produces a high initial (peak) pressure above the piston at combustion, but proportionately less pressure as the piston travels down the bore.

Setright's point is that the higher continuous pressure in the cylinder of a blown engine produces a more constant loading on the piston, rod, and crank during the power stroke than that in an unblown engine; and the same is true during the intake stroke when the blower is helping to pump air into the cylinder. Consequently the supercharger actually helps to reduce mechanical stresses on engine parts by "cushioning" inertia loads. Further, since the blower makes more power at given rpm levels, this means that a blown engine does not have to spin as tight to make the same power as an unblown motor, thus further reducing stress on engine parts.

EXPANSION RATIO AND THERMAL EFFICIENCY

A negative effect of the lowered static compression ratio necessary in a supercharged engine is a consequent lowering of the expansion ratio and thermal efficiency of the engine. The expansion ratio in a cylinder is a measure of how much the compressed air/fuel mixture in the combustion chamber can expand (and therefore exert pressure on the piston) before the exhaust valve opens somewhere near BDC on the power stroke and "blows down" the cylinder. It is the expansion of the heated gases in the cylinder that does the work to run the engine, so the expansion ratio obviously has a direct bearing on engine power.

If valve timing remains the same, the expansion ratio in an engine is proportional to the geometric (static) compression ratio. By lowering the

As we saw, performance auto designers were quick to learn the effectiveness of cooling the outlet charge of the supercharger, as evidenced by all the convection fins on these early blowers and manifolds.

compression ratio in a supercharged motor (that is, by enlarging the combustion chamber volume), we can get a larger quantity of air and fuel into the cylinder, which will expand with a proportionally greater force once it is burned (heated). However, considering that the cylinder is still the same size, this heated charge of gas will still have only the same distance to push the piston, and the same piston top area to push on. In other words, although we have greatly increased the force pushing on the piston for the entire length of its power stroke, the piston reaches the bottom of that stroke well before the power of the expanding gas has been used up.

Remember that the temperature of a gas drops as that gas expands in volume. In an engine cylinder, the heat which makes the gas expand in the first place comes from the energy content of the fuel (energy which can be converted to heat, and which can be measured in BTU's). The thermal efficiency of an engine is a measure of the percentage of potential heat energy in the fuel which is actually converted to work in the cylinders. Without going into a lengthy explanation of all these processes, I hope you begin to understand how the static compression ratio, expansion ratio, thermal efficiency, and ultimately the fuel economy of an engine are all closely related. In terms of supercharging, the results are that for a large increase in power, we must waste some of the potential power available, which means wasting some fuel, and this wasted power will show up as an increase in exhaust heat and/or water temperatures.

In a street supercharged engine the expansion ratio can be increased somewhat by retarding the exhaust valve opening timing and reducing intake/exhaust valve timing overlap. The extra pressure (heat) of the burned gases in the supercharged engine at the end of the power stroke will help push the exhaust through the valve and out the port.

SO LET'S GET PRACTICAL

The reason we have spent so much time discussing blower and engine theory was to answer, or at least begin to answer, some questions for those of you who need to know what makes things tick and why. Such questions are far too seldom asked about superchargers, and supercharging theory in general has progressed very little since the 1930's. We would also like to dispel certain misbeliefs about blowers.

But in this book—and in hot rodding in general—we are much more interested in practical application than we are in detached theory.

We know from practical experience that blowers work. We have given several reasons why they are not perfect, and have shown some ways in which they can be improved, but these are usually relatively minor faults. In most applications, the blower gives much more than it takes. In typical installations, a supercharger gives more power per cubic inch, more power per rpm, and more power per dollar than all the standard unblown engine modifications. On the other hand, do not expect a blower to give you better gas mileage, 'cause that's not part of the deal.

We have seen that the major problem associated with a street blower is heat. When the blower is making boost, it will be making more heat in the engine, and this means that you have to increase the capacity of the cooling system to carry away this heat before it damages engine parts. The heat carried away by coolant, or blown out the exhaust, is power wasted. Likewise, the more the blower heats the air in the intake manifold, the less efficient it is, the less potential power it will be pumping into the engine, and the more power it will be taking from the engine to operate.

Therefore, the most important and effective thing you can do to improve blower performance and efficiency is to keep the intake charge as cool as possible. This means drawing in cool air to the blower, using as efficient a blower as possible to keep manifold temperature down, and thereby lowering combustion temperature so that more of the power of the expanding gases can be extracted in the cylinder. Since we cannot very well raise the operating temperature of the engine (especially on a street motor), this means that the only way we can maximize engine power and efficiency, from the thermodynamic standpoint, is to lower the temperatures of all processes down the line *ahead* of the exhaust stroke. Lowering the intake and combustion temperatures will also lessen the tendency of the engine to knock on available gasoline.

In order to determine exactly how well the supercharger is working, or how much you are cutting down on intake heat, I strongly recommend installing both temperature and pressure sensors in the intake manifold. Without a good boost gauge and manifold temperature gauge in your car, you're driving blind with a blower.

Finally, we should realize that the big difference between street supercharging and competition supercharging is that a street machine is driven at part throttle the vast majority of the time. A gasoline engine is terribly inefficient in the first place, but it's considerably worse at anything less than wide open throttle. At part throttle the blower is basically a parasite on the engine (not to mention that a blower like a 6-71 GMC, plus drive, plus carbs, adds at least a hundred pounds of weight the engine must pull). Since it will not be making boost, or pumping "thick" air, a blower will take much less horsepower to operate at part throttle than it will at full throttle; plus manifold vacuum will actually spin a Roots-type blower like a windmill. But at part throttle a supercharger will very likely be taking more power than it is giving—not much, but some.

But most of us who want to bolt a blower onto our street motors aren't very worried about fuel economy in the first place, nor about part throttle engine operation. In most of the literature which has been written about blowers either in magazines or in books, much has been made about the "drawbacks" of blower noise and size. Several of the current blower manufacturers have gone to great lengths to fit superchargers under the hoods of our modern low-profile cars—usually compromising blower efficiency to do so. It is my opinion, and observation, that the majority of today's blower buyers have no intention of hiding the thing under a hood. Nor do they want a blower that nobody can hear. (In all honesty, I think this is one of the major deterrents of turbochargers for street or competition—not only do they make little noise of their own, but they also muffle exhaust noise; not to mention the fact that they just don't look good.) A big part of the appeal of supercharging is the sight and sound of the total package—added to the sudden kick in the pants only a supercharger can give the instant you mash the throttle. Furthermore, the majority of supercharged machines are hobby or special interest cars used primarily on weekends or just for fun cruising—much the way you'd use a ski boat or dune buggy—rather than daily transportation vehicles.

On the other hand, if you really are looking for a supercharger to install in a practical way to run every day on your transportation car, for instance to give a small displacement 4-cylinder more pep when it needs it, there is no reason why such an application cannot be trouble free, unobtrusive, quiet, and effective. Several new superchargers, especially smaller ones, have recently been marketed primarily for such applications, and we will cover them in detail. So let's see what the market has to offer.

Recent blower builders have tried to devise ways to cool the blower charge, but it isn't easy. A water-cooled GMC-type case, like this, obviously needs a cold water supply to begin with (such as in a boat), but still isn't very effective since most of the heat is being made in the manifold, where the air is being compressed.

Gale Banks, who once ran a very successful Bonneville Corvette using an intercoller *without* a blower (which still produces the effect of a supercharger by increasing the density of the intake air), holds an example of a large, heavy-duty industrial heat exchanger which could very well be mounted under a healthy GMC blower...provided you have something very cold to run through it.

Blown gas drag boat racer Bob Brownell is one of the few to actually devise an intercooler for a GMC type blower. The system on his boat picks up cold lake water, circulates it through coils inside a small tank (2) packed with dry ice, and then through a large Harrison heater core in an aluminum spacer below the blower (1). The system proved good for 100 hp on the dyno. The only problem, says Bob, is the strength of the heater element—the blower tends to tear the fins out.

CHAPTER 5

THE ROOTS SUPERCHARGER

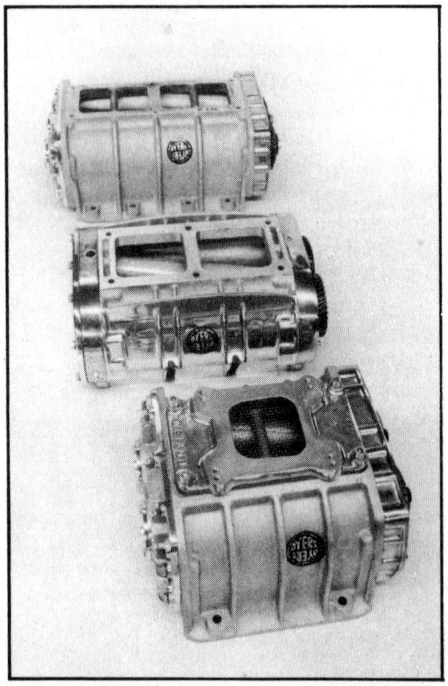

Although there are several types and sizes of GMC Roots type blowers, the ones most commonly used for automotive supercharging are the 4-71, 6V-71, and 6-71, as shown front to back in this line-up from Supercharger Engineering. This company, however, is the only one currently adapting the plentiful 6V-71 blowers to cars.

PART 1: "JIMMIES"

Although we have seen that the Roots Brothers air blower was invented in the 1800's, and that Roots-type blowers were used extensively for automotive supercharging in Europe before World War II, it is the GMC Roots-type blower that brought supercharging to hot rodding. No other piece says "blown" like a big, polished 6-71 on top of a V-8 engine.

Rodders and racers are most familiar with the 71-series of GMC Roots-type blowers. These were the first GM blowers, designed in the early 1930's by the GM Research lab and introduced on the 6-71 engine, General Motors' first diesel. Later the Detroit Diesel division of GM was formed, and the blowers and diesel engines are now produced by the Detroit Diesel/Allison division of General Motors.

The numeric designation "6-71" refers to a 6-cylinder, inline, diesel engine, each cylinder measuring 71 cubic inches, for a total displacement of 426 cubic inches. The 71-series of GM diesels included 2, 3, 4, and 6 cylinder inline engines, each with its own size of blower, known by the same numeric designation as the engine, i.e., 3-71, 4-71, 6-71. These blowers all use the same basic design and the same rotor diameter; just the length of the case and rotors was changed to increase or decrease the blower's pumping capacity.

Several other types and varieties of GM blowers have been produced over the years. The 53-series, for instance, uses straight-cut 2-lobe rotors rather than helical 3-lobe rotors. This GM design has been copied by some of the current aftermarket blower manufacturers discussed later. The 53-series blowers are used on both inline diesels (i.e.,3-53, 4-53) and Vee engines (6V-53, 8V-53). There are also 71-series blowers for Vee engines, but they differ in both size and shape from inline 71 blowers. They come in two basic sizes, the 6V-71 and the 8V-71, which use 3-lobe helical rotors of similar design, but smaller size, than inline 71-series GMC's. These blowers are also used in pairs for 12V-71 and 16V-71 engines, and at higher overdrive ratios (approximately 2.5:1 instead of approximately 2:1) for 92-series engines. The V-71 series blowers are quite common and plentiful these days, but only one automotive blower manufacturer—Gary Dyer's Supercharger Engineering—has designed kits and components to use 6V-71's on automotive engines.

Each of the GMC blowers is used in combination these days with turbochargers, intercoolers, and aftercoolers for a wide variety of engine applications and designations; but the blowers themselves, other than differences in seals and overdrive ratios, are basically the same in each series. However, there is one critical exception. In approximately 1978, GMC introduced a new line of 71-series blowers, labeled simply "4-71" and "6-71," which have smaller diameter rotors than previous 4-71's and 6-71's. (The reason for this reduction was to reduce horsepower required to drive the blower, to reduce fuel consumption.) The difference in rotor size is hardly perceptible, even with a pair of these blowers side by side. The exterior of the cases, and the end plates, gears, etc., are the same. However, the large (early) rotors will not fit in the small (late) case; on the other hand, the small rotors *will* fit in the large case, but will give clearances that would make the blower nearly ineffective. Either size of 4-71 or 6-71 could be used for an automotive performance blower, with aftermarket end plates, drives, manifolds, etc., as long as the rotors match the case, and are clearanced properly.

Finally, each of the inline GMC blowers is available in either right-hand or left-hand, front-drive or rear-drive configuration. In other words, if you were to buy an inline Detroit Diesel engine, you could order it with the power take-off at the front or rear (that is, with clockwise or counterclockwise crankshaft rotation), and with the blower mounted on the right or left. This is to allow versatility in mounting and maintaining these engines, either singly, or in pairs (as in a boat, where twin props spin in opposite directions). Consequently a GMC blower straight off a diesel might turn "frontwards" or "backwards" on top of an automotive V-8. Remember that auto engines turn clockwise (viewed from the front), and that the blower rotors should turn towards the outside of the case when viewed from the top. If the blower turns "backwards," the rotors can be flopped side for side, and the rotor shafts changed end for end, to make it work properly on a car. Once the "tang" or "lip" has been machined off the bottom of a GMC case to mount it on an automotive manifold, the case itself has no front or back.

The 71-series GM blowers are the ones of particular interest to hot rodders—primarily the 4-71 and the 6-71, since these are the two sizes most often adapted to automotive V-8's. These blowers were designed to scavenge exhaust and blow fresh intake

Hot rodders are most familiar with the 71 "inline" series of GMC blowers, which have appeared in slightly different sizes and forms since their introduction in the late 1930's. From left to right we see a 6-71, 4-71, 3-71, and 2-71 ready to mount on Jimmy diesels.

A unique characteristic of 71 series GMC blowers (both inline and V types) is the three-lobe, "twisted" or helical rotors. The twist in the rotors reduces pulses in the manifold, but does cause some problems in high-boost racing applications.

Despite several new varieties on the market, GMC 6-71 and 4-71 blowers are by far the most plentiful on hot rods. The 6-71, at left, has later-style end plates and covers, while the 4-71 has the earlier, thicker bearing plates and covers.

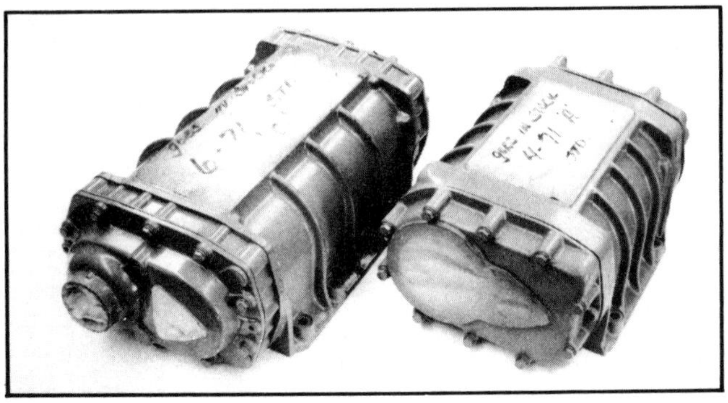

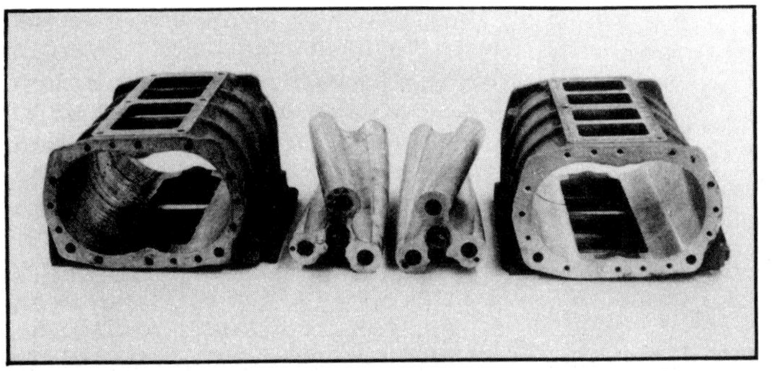

The difference between a standard size 71-series blower (left) and a newer "small diameter" blower (right) is barely perceptible, and would make little difference in performance on a street engine. However, you don't want to mix these rotors and cases.

V-series GMC blowers are used on V-type engines rather than inlines, and are usually used in conjunction with a turbocharger these days. Left to right, this is an 8V-92, an 8V-71, and a 6V-71.

Inline GMC blowers hang on the side of big 2-stroke diesels and blow dry air directly into the cylinder ports. This is a 6-71 engine—a 426 cubic inch six cylinder.

V-series blowers are used on V-type 2-stroke diesels, and most are dual-staged with a turbocharger these days, as shown on this 8V-71 engine—a 568 cubic inch V-8.

air into 2-stroke diesel engines (since a 2-stroke doesn't have separate intake and exhaust strokes, it needs help getting the exhaust out and the fresh air in on the same stroke). Specifically, on the diesel the blower mounts to the side of the engine, with its outlet positioned over the cylinder intake ports. When the piston travels down the bore to uncover the intake ports near the bottom of the cylinder, the blower pumps fresh air in at the bottom while the exhaust valve opens at the top. Thus the fresh air pumped in by the blower pushes out the exhaust as it fills the cylinder.

As you can see, the GM blower is designed to pump dry air, to turn at relatively slow speeds, and to operate at zero boost levels. This is why GM calls these units "blowers" rather than superchargers. In the diesel their function is simply to *move* air, not to compress it, since the exhaust valve opens as soon as the blower starts pumping air into the cylinder.

This is where a Roots-type air pump shines—it is excellent for moving very large quantities of air at relatively slow rotor speeds against low backpressure resistance. Since it is a positive displacement pump, it moves a given quantity of air with each revolution, which means that it has immediate response.

The fact that Jimmies move very large quantities of air, especially at lower rpm's, relative to other types of blowers, is what makes them so good for supercharging automotive engines, especially those of 300 cubic inches or more. However, when we bolt a GMC atop a 4-stroke V-8 and overdrive it to make boost, we are asking it to do something for which it was not intended.

As is so often the case with automotive engines, especially of the high performance variety, we end up making lots of minor compromises in order to effect a big overall gain.

In the case of the GMC blower, its volumetric, mechanical, and adiabatic efficiencies are all highest when it is pumping air against no load (that is, making zero boost), as it was intended to do. As soon as we begin to use it to compress air, all three of its efficiency ratings go down. Obviously it will take more power from the engine to turn the blower against backpressure, thus lowering the mechanical efficiency as boost goes up. Also, as backpressure increases, this will force more air to leak through the clearances between the rotors and the case, lowering volumetric efficiency. And the more work that has to be put into the blower to turn it, as well as the more backflow leakage, the more heat will be transferred to the air, thus lowering the adiabatic efficiency.

Furthermore the Roots design—which is excellent for pumping—is just not a very efficient way to compress air. The Roots is a backflow compressor, which means that it compresses air by

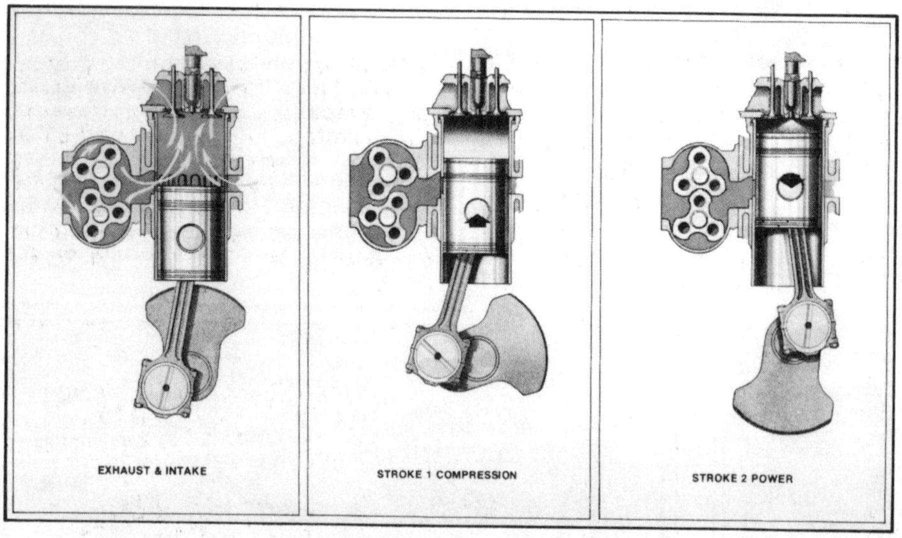

Here's how a Jimmy was designed to work. On the compression stroke the 2-stroke diesel compresses dry air blown into the cylinder by the blower. When the piston reaches the top of the cylinder, the fuel injection nozzle sprays a timed amount of fuel into the chamber and it ignites spontaneously from heat. The piston is then forced down on the power stroke until it uncovers a ring of holes (ports) in the cylinder wall near the bottom, at which time the exhaust valves (usually two) open. With the ports uncovered by the piston at the bottom of the bore, and the exhaust valves open at the top, the blower pumps in a shot of fresh air, which forces the exhaust gases out ahead of it. Ingenious, isn't it?

The 53-series GMC blowers have been around for years, but have largely been ignored by automotive enthusiasts until recently (both the Penco and the B&M blowers are basically copies of the GM 4-53). This line-up includes, right to left, a 2-53, 3-53, 4-53, 6-53, and 8-53.

cramming it into a container (the manifold) outside the blower. Imagine a Roots with straight cut rotors for a minute. As the rotors turn past the blower intake opening, they trap a certain volume of air in the cavities between the lobes and carry that air around to the outlet opening. If there were no pressure in the manifold, the Roots would simply pump the air out the other side. But when the pressure on the outlet side is much greater than that on the intake side of the blower (that is, when there is boost in the manifold), as soon as the rotor tip passes the edge of the outlet opening, manifold pressure will force air to rush backwards into the rotor cavity, compressing the air trapped inside—thus the term backflow compressor. As you can imagine, this isn't the smoothest way to compress air, and this continuous back-and-forth motion adds greatly to air turbulence and friction, which in turn heats the air, lowering the adiabatic efficiency of the blower. As one supercharger expert put it, "The biggest problem with the Roots is that it beats the (expletive) out of the air."

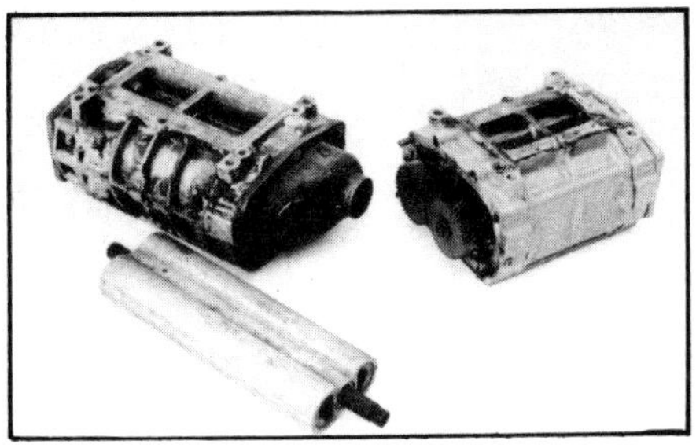

The major difference between 71-series and 53-series GM blowers, other than the obvious fact that the latter are made for smaller motors, is that the 53's use straight-cut, 2-lobe rotors.

In the 71-series GMC's, however, the rotors are twisted (helical cut rather than straight cut). This serves two purposes in the stock blower: it reduces blower noise (since the air backflowing into the cavity produces a sort of shock wave), and more importantly it reduces pulsing in the output airflow. When a GMC is used to compress air on an automotive engine, the rotor twist helps reduce the abruptness of the backflow compression process, making the job of compressing the air "smoother" and increasing adiabatic efficiency somewhat.

The fact that 71-series GM blowers have three rotors instead of two also greatly reduces airflow pulsing, increases blower smoothness, and helps to reduce leakage back through the blower, since more than one rotor tip will be sealing against the case walls at the same time (that is, there will be more than one rotor tip per revolution making a seal between manifold pressure and inlet pressure).

Not much research has been devoted to the effect of supercharger pulsing on engine performance. It stands to reason that if the blower is pumping air in pulses or waves, and those waves do not coincide with the opening of intake valves, some cylinders might get a stronger charge than others. Or, if the intake valve were timed to open just when the manifold air mass was backflowing into the rotor cavity, it would get a weaker "shot" from the blower. In other words, you would want to time the blower so that intake valves opened at peak pulses rather than at "trough" or "ebb" pulses. Supposedly some of the pre-war Mercedes superchargers, most of which were of the 2-lobe design, were actually timed to the engine in this manner. But this would be very difficult to devise. Increasing the number of ro-

Underside view of a 6-71 shows how the helical rotors mesh. As they carry air/fuel mix around the outsides of the case, the "propeller" action of the twisted rotors also moves it toward the front of the blower (which is on the right in this picture). Arrows point to small holes in stock GMC case which allow oil to drain out of bearing plates on diesels. If you use stock end plates, you must plug these holes to retain oil in the gear case. Non-GM bearing plates do not have these holes.

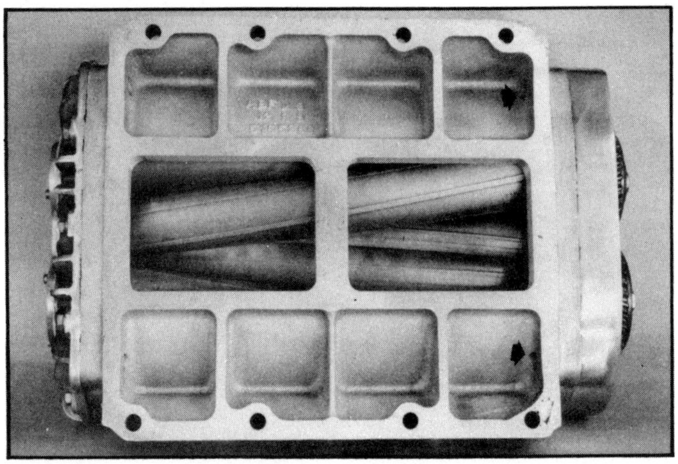

GMC blower builders have fought rotor twist problems for years on high performance race cars, but the cost of making custom rotors is usually worse than the problem. However, a few manufacturers have experimented with 2-lobe, straight-cut rotors for GMC's. The ones shown here are made by Don Hampton, and have shown good results on the dyno.

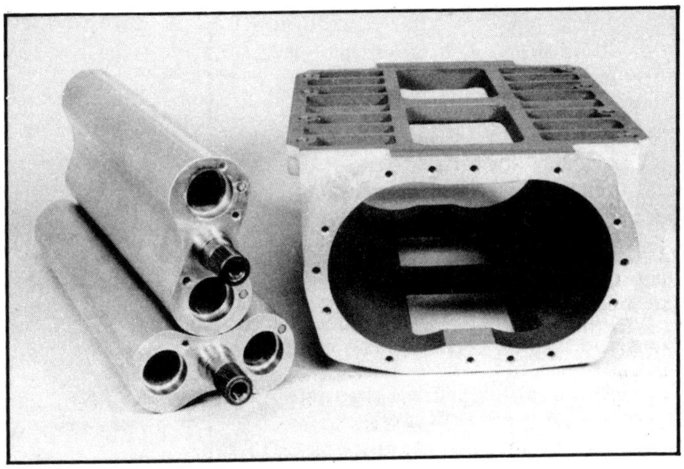

tors, and twisting them, are much easier solutions. This is one point to bear in mind when comparing GMC's with other Roots blowers.

The major reason why other blower manufacturers do not use twisted rotors, or why they use a 2-lobe rather than a 3-lobe design, is to reduce production cost. Machining helical rotors is difficult and extremely expensive. For the same reason, nearly all GMC blower rebuilders use factory GM rotors, even in racing blowers. However the twist in GM rotors can also cause some problems. In high rpm, high boost applications the GM rotors will tend to "untwist" slightly as they turn against a load. This can cause galling of the rotors if they actually touch, and it also requires wider than optimum rotor-to-rotor clearance. Second, the spiral GMC rotors will force *air* to the front of the blower, while the heavier *fuel* will cling to the rotor surfaces—or may be forced rearward as the car accelerates—causing the front cylinders to run lean and the rear cylinders to run rich.

However, all of these GMC "problems" are greatly diminished in a street application where the blower will be making less than 10 pounds of boost, will be driven at crankshaft speed or less, and will usually be turning less than 5000-6000 rpm.

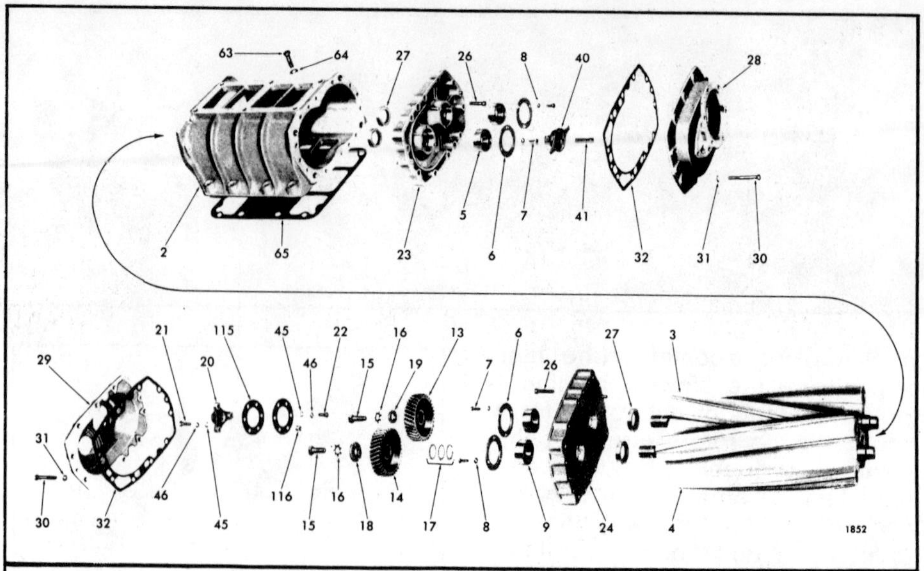

2. Housing--Blower
3. Rotor--Blower--Upper--R.H. Helix
4. Rotor--Blower--Lower--L.H. Helix
5. Bearing (Roller)--Front
6. Retainer--Bearing
7. Bolt--Bearing--Retainer
8. Lock Washer
9. Bearing (Ball)--Rear Double Row Thrust
13. Gear--Rotor--Upper--R.H. Helix
14. Gear--Rotor--Lower--L.H. Helix
15. Bolt--Rotor Gear
16. Lock Washer--Bolt to Rotor Gear
17. Shim--Gear to Bearing (For Timing Rotors)
18. Disc--Fuel Pump Coupling
19. Washer--Rotor Gear Retaining
20. Hub--Rotor Drive
21. Bolt--Plate to Gear
22. Bolt--Plate to Hub
23. End Plate--Front
24. End Plate--Rear
25. Pin (Dowel)--Housing to End Plate
26. Bolt--End Plate
27. Oil Seal--End Plate
28. Cover--End Plate--Front
29. Cover--End Plate Rear
30. Bolt--End Plate Cover
31. Lock Washer
32. Gasket--End Plate Cover
40. Coupling Assy.--Water Pump Drive
41. Bolt
45. Plain Washer
46. Lock Washer
63. Bolt--Blower Mounting
64. Plain Washer--Blower Mounting
115. Plate--Blower Rotor Drive Hub
116. Spacer--Place to Gear

A standard shop manual exploded view of a 6-71 GMC shows relative placement of parts. It also graphically demonstrates the utter simplicity of this piece of machinery.

STRENGTHENING THE GMC BLOWER

To convert a GM diesel blower for street or performance use on an automotive engine, two things need to be done: it must be strengthened in certain areas to withstand higher rotational speeds and higher stresses from pumping against pressure; and the clearances must be reset. You can think of these operations as blueprinting the blower, just as you would blueprint an engine in much the same manner for performance use. We will describe GMC blower blueprinting in this book primarily so that you can see and understand the several operations involved. However, rebuilding a Jimmy requires a considerable amount of machining, access to the proper components and processes (such as hard anodizing), and an experienced eye and feel for just the right tolerances. We do not recommend that

The chart at right gives all the pertinent dimensions of 71-series GMC blowers—figures which few blower builders have ever seen. The blower drive ratio listed is that used on the GM diesels.

CHARACTERISTICS OF GMC BLOWERS (SERIES L-71)

ENGINE	BLOWER DRIVE RATIO	ROTOR DIAMETER INCHES	ROTOR LENGTH INCHES	BLOWER DISPLACEMENT CU.IN./REV.	DISPLACEMENT PER INCH OF ROTOR LENGTH
2-71	2:1	5.142	6.110	136.0	22.2
3-71	2:1	5.778	7.482	205.38	27.45
SMALL DIAMETER ROTOR BLOWER ASSEMBLIES					
4-71	2:1	5.502	9.980	225.94	22.64
4-71T	1.95:1	5.505	9.980	225.94	22.64
6-71	2:1	5.505	14.975	339.03	22.64
6-71T	1.95:1	5.505	14.975	339.03	22.64
INLINE 2-VALVE					
4-71	2:1	5.778	9.980	273.95	27.45
6-71	2:1	5.778	14.975	411.06	27.45
V-SERIES ENGINES					
6V-71	2.05:1	5.778	11.920	327.20	27.45
6V-71T	1.95:1	5.778	11.920	327.20	27.45
12V-71	2:1	5.778	(2) 6V Blwr	654.40	27.45
12V-71T	1.95:1	5.778	(2) 6VT	654.40	27.45
8V-71	2:1	5.778	15.894	436.29	27.45
8V-71T	1.95:1	5.778	15.894	436.29	27.45
8V-71P	2.05:1	5.778	15.894	436.15	27.45
16V-71	2.05:1	5.778	(2) 8V Blwr	872.58	27.45
16V-71T	1.95:1	5.778	(2) 8VT	872.58	27.45

A GMC blower is a relatively simple device with few moving parts, and it is designed for very rugged use on commercial engines. This photo shows the basic components, and a few modifications which will make a stock blower strong enough for most street applications: A & B—stock rotors showing tips where rotor-to-case clearance is checked; C—bearing plate alignment dowel holes, which position rotors in case; D & E—holes drilled to spin-balance each rotor; F—steel rotor shaft, which is pressed into end of hollow rotor, then pinned; G—stock drive gears; H—steel reinforcing rings around bearing boss in stock end plate; I—high-quality double-row ball bearings.

Although most competition blower builders make their own cases these days, a stock GM case is fine for street if it's in good condition. These are two styles of 6-71 cases; the older one on the left has a full lip on one side where it hangs on the diesel motor (this must be machined off for automotive applications), plus has smaller inlet ribs than the newer model. Also note that the older case is badly scored inside, making it unfit for use.

you try to convert a stock blower to street performance specs by yourself. There are plenty of excellent blower builders who are set up to do the job efficiently, quickly, and correctly.

A Roots blower is a simple machine. It has only two major moving parts, the rotors; and these should never touch, so they should never wear out. The only parts that should ever suffer any wear in a GMC blower are the bearings and the seals. Since the bearings hold the rotors in position—front to back, side to side, and up and down—they are obviously critical to maintaining close rotor tolerances, which will be set to within .001-inch. The stock GM rear roller bearings are too sloppy, and should be replaced with better ball bearings. The front stock double row ball bearings should also be upgraded. Likewise, the stock rotor shaft oil seals should be changed for good neoprene seals, which will keep gear oil out of the blower, and blower pressure out of the gear case.

For any street blower the stock GMC rotors and case will work fine. In fact, new GM rotors are very expensive, so the vast majority of 4-71's and 6-71's that are set up for hot rods are rebuilt from used blowers. As long as the bearings did not wear out, allowing the rotors to rub the case or each other, and as long as no foreign matter had been ingested by the blower in the past (dirt, rocks, etc.), a used case and rotors should be fine. Even small scratches inside the case are not a problem in a street blower; they can be lightly hand sanded. A good first step in rebuilding a GMC for street performance use, however, is to have both the inside of the case and the outside of the rotors hard anodized. This process will not only toughen the surfaces of these parts, but will also increase surface thickness approximately .001-inch, which will help "tighten up" the clearances between rotors and case. Remember that air can erode soft metal surfaces like aluminum—especially air that has any dirt or dust particles in it. The anodizing will reduce surface erosion in the future at the same time that it will replace some of the surface worn off in the past. And *always* use a good air cleaner on a street blower setup.

Since stock GMC rotors are hollow inside and the shafts are stubs inserted in each end and held in place with single pins, most rebuilders double-pin the shafts for performance applications. This might not be mandatory on street blowers, especially 4-71's, but it is a simple procedure and there's little reason not to do it.

A section cut from the middle of a GMC 71-series rotor shows its hollow construction.

On a street blower you could also use the stock GMC end plates (bearing retainer plates), which are the same for front and rear. If you do, you should flycut the outside of the bearing bosses and press steel support rings in place to strengthen this area, which is prone to cracking when the blower is run at high rpm and high boost. Also, the stock end plates have small holes in the bottom corners to allow oil to flow from one end plate to the other through the housing. These holes must be plugged (with small freeze plugs) for automotive applications. However, most blower rebuilders make their own end plates for GMC's which have stronger bearing support bosses and ribs cast in, which eliminate oil holes, and which (for the rear) are considerably thinner than stock plates to allow more room for a distributor or magneto on most V-8's. These custom end plates are not very expensive (about $50 each, new), and would be good insurance on a street blower. When you are having the case and rotors hard anodized, also have the inner surfaces of the end plates done as well.

Early GMC's used cast iron gears which are about one inch thick. Later versions use steel gears which are about 3/4-inch thick and have either 30° or 45° helical cut teeth. The steel gears are definitely stronger, and are necessary on a race blower. Most

Even race blowers use stock GMC drive gears. The one on the right is an older cast iron type, while the one on the left is the thinner, preferred, steel gear with 30 degree cut teeth.

rebuilders recommend them for street units, too. The helical cut of the gear teeth reduces blower "whine," but when the blower is driven at high speed, the angled teeth want to separate, forcing one rotor forward and the other back. This changes the clearance between the rotors, and can sometimes cause galling. On a street blower this problem is not critical, but the 30° gears are universally preferred.

Finally, late model GMC's use a flex type drive hub (which connects the blower drive shaft to the left gear); it looks sort of like a steering coupler, and it tends to fall apart under stress. Use an earlier solid hub, or a custom one from a rebuilder. As an interesting note, one reason for the late GM coupler was to reduce the transmission of engine vibrations from the crankshaft to the blower in stock diesels. GM also used a torsion bar blower drive shaft for this reason. If you were contemplating any sort of direct drive between the crankshaft and the blower on your street machine, similar to the early front-mount drag racing layout, you might encounter blower damage from crank vibrations if used for extensive driving. However, current belt drives for top-mounted street blowers will absorb enough crank vibration to eliminate this problem.

You will probably be buying a complete drive assembly to adapt the GMC blower to your engine, and this should include a drive shaft, snout, and provision for a belt-tensioner idler pulley. On a street blower you can use the stock GMC front plate cover with no problem (stronger ones are made for race blowers), but you will need a cover plate for the right gear inspection hole, and you will have to add a vent and filler hole for front gear oil.

1/2" PITCH	TOP PULLEY 32	33	34	35	36	37	38	39	40	41	42	43
BOTTOM PULLEY 32	0	3%	6%	9%	13%	16%	19%	22%	25%	28%	31%	34%
33	3%	0	3%	6%	9%	12%	15%	18%	21%	24%	27%	30%
34	6%	3%	0	3%	6%	9%	12%	15%	18%	21%	24%	26%
35	9%	6%	3%	0	3%	6%	9%	11%	14%	17%	20%	23%
36	13%	9%	6%	3%	0	3%	6%	8%	11%	14%	17%	19%
37	16%	12%	9%	6%	3%	0	3%	5%	8%	11%	14%	16%
38	19%	15%	12%	9%	6%	3%	0	3%	5%	8%	11%	13%
39	22%	18%	15%	11%	8%	5%	3%	0	3%	5%	8%	10%
40	25%	21%	18%	14%	11%	8%	5%	3%	0	3%	5%	8%
41	28%	24%	21%	17%	14%	11%	8%	5%	3%	0	2%	5%
42	31%	27%	24%	20%	17%	14%	11%	8%	5%	2%	0	2%
43	34%	30%	26%	23%	19%	16%	13%	10%	8%	5%	2%	0

FIGURES IN WHITE AREA ARE OVERDRIVE PERCENTAGES.
FIGURES IN SHADED AREA ARE UNDERDRIVE PERCENTAGES.

These charts are very handy for figuring blower overdrive or underdrive with given sets of pulleys. On the 8mm pitch chart, only overdrive ratios are given, but swapping the pulleys, top for bottom, will give the same percentage in underdrive. To change overdrive or underdrive percentage without having to change to a longer or shorter blower belt, choose a set of pulleys with the same combined total number of teeth as the pulleys on the motor.

8 MM PITCH	TOP PULLEY 52	51	50	49	48	47	46
BOTTOM PULLEY 63	21.1	23.5	26.0	28.5	31.2	34.0	36.9
64	23.0	25.4	28.0	30.6	33.3	36.1	39.1
65	25.0	27.4	30.0	32.6	35.4	38.2	41.3
66	26.9	29.4	32.0	34.6	37.5	40.4	43.4
67	28.8	31.3	34.0	36.7	39.5	42.5	45.6
68	30.7	33.3	36.0	38.7	41.6	44.6	47.8
69	32.6	35.2	38.0	40.8	43.7	46.8	50.0

SETTING THE CLEARANCES ON A GMC

Strengthening a Jimmy for performance automotive use is relatively easy. The tricky and tedious part is setting the clearances around the rotors. The air gap around the rotors will significantly alter both the volumetric and the adiabatic efficiency of the blower, as well as determine whether the blower will make boost or destroy itself in traffic. The problem is that the blower is not "adjustable." As I said, clearancing a blower is very much like blueprinting a motor: you've got to set clearances a little looser for high performance in a race motor, but not so loose that the pistons won't seal or the rods will knock; and to change clearances you'll have to custom machine the components after careful measuring. The procedure is the same in a supercharger.

In an accompanying photo section we will show you the various steps taken by one of the best blower rebuilders, Gene Mooneyham, to set up a 4-71 for the street. Without going into detail, the basic operation is to set the clearance between the back of the rotors and the face of the rear plate, between the front of the rotors and the back of the front plate, between the edge of the rotor tip and the case wall at the outlet port, and less importantly, between the rotor tip and the case wall at the inlet port. The front and back clearances are set by machining the ends of the case and the rotors, and the rotor-to-case clearances are changed by realigning the end plates on the case, and then redrilling the end plate alignment dowel locations and inserting larger dowels.

Also reset is the rotor-to-rotor clearance "up the middle" of the blower. Since it is nearly impossible to move the rotors closer together or farther apart, the goal is to set the timing of the twisted rotors so that the clearance between any two rotor lobes is the same (not accounting for any rotor "untwist" or deflection at speed...which should not be significant in a street blower). The timing of the rotors is controlled by the meshing of the drive gears. Mooneyham likes to swap different gears to get the right tolerances (he has found as much as .006-inch clearance difference just by changing gears). Otherwise the gears can be machined on the back (every .003-inch machined from the back of the gear changes rotor timing .001-inch), or shims can be inserted behind the gears—although this last option is the least preferable.

Most blower rebuilders feel that the rotor-to-case clearance at the bottom of the blower and the front and back rotor end-clearances are the most critical, followed by the rotor-to-rotor clearance in the middle, and then the rotor-to-case clearance at the top. Obviously, they must make compromises given the limits of which clearances can be feasibly altered.

Since there is no practical way to tighten up all of the blower clearances at once, other than the slight surface increase afforded by hard anodizing, the only other alternative for an overly "loose" blower is to machine the rotors to accept small, half-round strips of Teflon, which are cut down to protrude about .002-inch from the rotors. Since the Teflon can come in much closer contact with the aluminum before damaging it, clearances can obviously be set much tighter for a more efficient blower. This has become the standard practice on all drag racing GMC type superchargers.

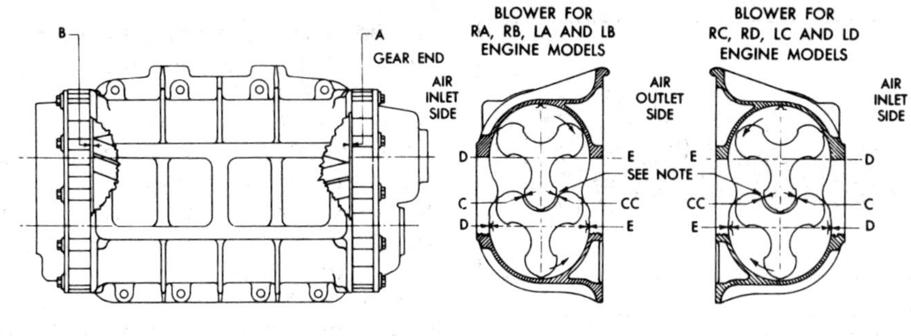

		A	B	C	CC	D	E
3-71	MIN.	.007	.007	.010	.004	.015	.004
	MAX.				.008		
4-71	MIN.	.007	.009	.010	.004	.015	.004
	MAX.				.008		
6-71	MIN.	.007	.014	.010	.004	.015	.004
	MAX.				.008		
6-71P	MIN.	.012	.019	.014	.002	.030	.004
	MAX.				.006		

TIME ROTORS TO DIMENSIONS ABOVE

NOTE: TIME ROTORS TO DIMENSION ON CHART FOR CLEARANCE BETWEEN TRAILING SIDE OF UPPER ROTOR AND LEADING SIDE OF LOWER ROTOR (CC) FROM BOTH OUTLET AND INLET SIDE OF BLOWER

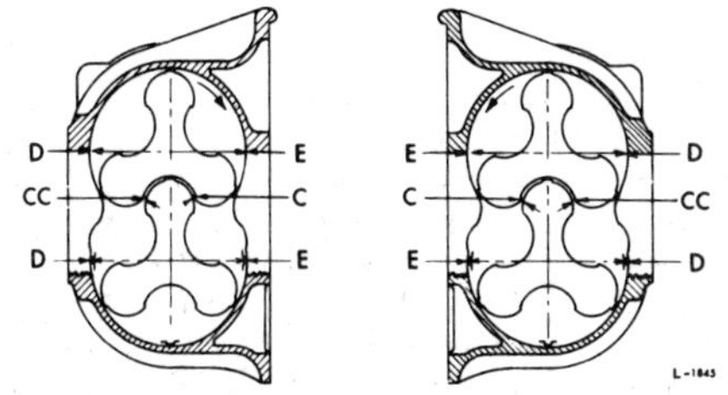

Chart of Minimum Clearances for Standard Blowers, Smaller Diameter Rotor Blowers and Large Bearing (P) Blowers

This diagram from a Detroit Diesel repair manual shows where blower clearances are measured and how much the clearances should be in a stock diesel unit. As you can see, a stock blower set to minimum clearances is pretty close to the specs used to set up a Jimmy for street use on an automobile engine.

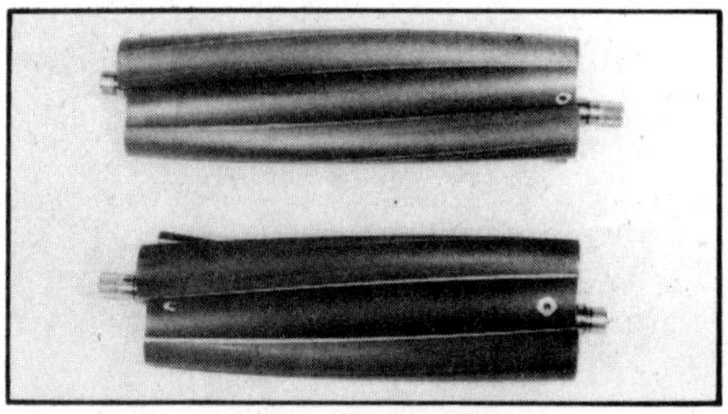

These rotors, prepared by Gene Mooneyham, have been hard anodized (note black tinge) and fitted with Nylatron strips both along the leading edges and on the inside curve where the rotors mesh in the middle of the blower. The upper rotor has been lengthened, by pressing on a piece of another 6-71 GMC rotor, to fit a 10-71 case.

However, opinions vary about using Teflon in street blowers. The big difference between a street blower and a race blower is that on a drag race machine, especially one running large quantities of cold fuels like nitro or alcohol, the blower never gets warm, let alone hot. Therefore little accommodation need be made in a race blower for heat expansion of the parts. On the other hand, race blowers are spun at high rpm under heavy loads, which causes both rotor growth from centrifugal force, and rotor untwist. In a street blower, on the other hand, rpm and boost levels will be lower, but heat will be a much greater factor, especially considering that the street motor will be running on gasoline, and you won't want to run it overly rich all the time, as you could on a race car.

This is the main reason why clearances are set slightly looser on a street blower than on a race blower. Incidentally, there is a slight amount of truth to the old wives' tale that a worn out drag blower makes a good street blower. But bolting a used blower onto your engine without checking all the clearances first is a lot like buying a used race motor at the swap meet and running it without ever pulling the heads or pan. Maybe you can take the seller's word for the condition of the piece, but you certainly won't know what's really inside unless you take it apart and measure it. Blower rebuilder Don Hampton states categorically, "I would never buy a used race blower." In a following section he will give you some pointers on how to shop for a used GMC supercharger.

But clearances are not the only factor. Race blowers these days usually use Nylon or Nylatron strips in the rotors. These should definitely not be used in a street blower, or in any gasoline supercharger since the heat will destroy them. Some builders even question the reliability of Teflon inserts in a street blower, claiming that the Teflon strips can get chewed up or will disintegrate in time, or else will quickly wear down in a street blower, leaving clearances too loose. Certainly a Teflon-lined blower will need checking and rebuilding more often than a blower without Teflon, but the Teflon-lined blower could be more efficient (thereby causing less heat rise) in the short run. The real question is how many miles do you really expect to put on your blown car, and how tight do you want the blower?

HOW TO BUILD A GMC STREET BLOWER

We certainly do not expect many of you to have the equipment, nor the machining experience, to be able to rebuild and clearance a GMC blower yourself—nor do we recommend it. There are several highly experienced and competent GMC blower rebuilders who will do the job for you at prices which are surprisingly reasonable considering the amount of work the job takes. Although most of the blower shops are located in Southern California, they all do business on a national, and even international, basis and have several means for shipping blowers across the country.

If you're a long-time drag racing fan, you undoubtedly remember the incredible Mooneyham and Sharp 170 mph, blown fuel, steel-bodied '34 Ford coupe which smoked the slicks all the way down the track in the early 60's. Gene Mooneyham has been building GMC's for at least 30 years, and is generally regarded as one of the best—and certainly one of the most meticulous—of current blower builders. In this photo section Gene, and his son "Little Gene," demonstrate all of the tedious steps required to convert a stock 4-71 GMC blower into a precision automotive street supercharger. We present this information not so that you can do the job yourself, but so that you can familiarize yourself with the processes involved, and appreciate how much work goes into a good GMC rebuild. Not all of the blower builders follow every step shown here, but this is how the job is done properly.

Gene Mooneyham spends most of his time building big, custom-case, Nylatron-lined competition blowers for racers, such as this 14-71. Such blowers require even more construction and clearancing steps, as well as more custom (non-GMC) parts than shown in the following photos. Unlike some of the race blower builders, however, Gene also welcomes street blower work.

Any GMC blower build-up must start with some basic GMC pieces. For a street blower, you can use the stock rotors, case, gears, front bearing plate, and front cover, and good used components, such as these, are just as good as new ones.

Gene points to side of case where lip has been cut off to allow the blower to sit flat on the manifold. Next the used case is flycut, as shown, on the top and bottom to make sure surfaces are flat and parallel.

Then the front and back of the case are each flycut, first to square up the surfaces, but also to "shorten" the case slightly in relation to the rotors, so that rotor-to-end plate clearance can be adjusted later. A total of about .010-inch is removed.

After this preliminary work the case is sent out to be hard anodized (note dark color on the inside), and polished if the customer desires. The inner surface of each end plate is also anodized, and in this case the outsides were also polished at this time.

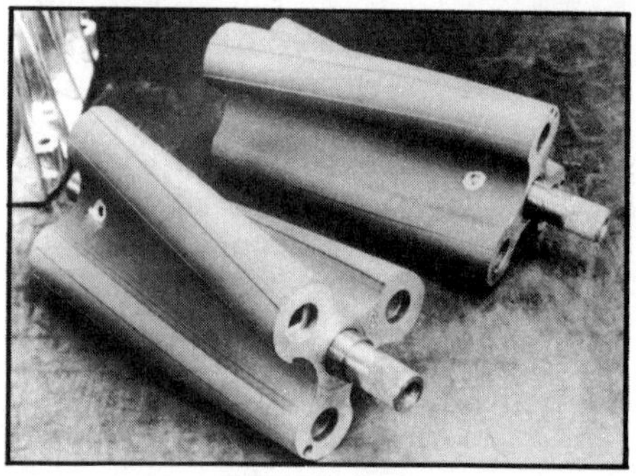

The stock rotors are also sent out for anodizing, which not only toughens the surface, but also slightly increases it to help tighten up clearances. The rotors and shaft stubs are also drilled, and an extra retaining pin is pressed in at each end.

This photo compares stock GM bearings with Mooneyham's replacements. Rear bearings are shown on the left, and a ball bearing (right) replaces the stock roller bearing (left). At the front, the stock double-row ball bearing (left in right-hand pair) is replaced with one which has a thicker inner race against which the drive gear rides.

The new bearings are pressed into the rear bearing plate, in this case a Mooneyham-made part.

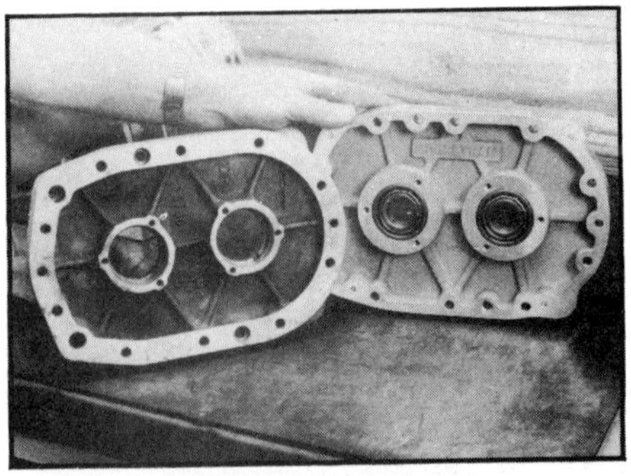

Although a stock front bearing plate could be used on a street blower, this customer preferred a beefier Mooneyham plate (right) as used in race blowers. The increased bearing support is obvious. Front bearings have been pressed in place.

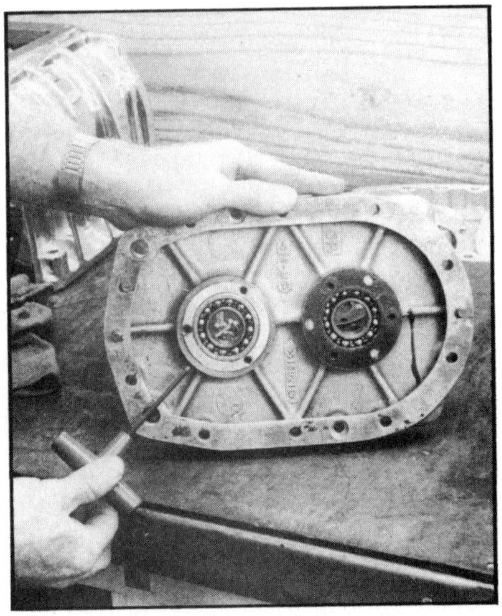

If a stock bearing plate were used, Gene would turn down the outside of the bearing boss slightly and press on steel support rings, as shown.

The rotor shafts have a friction fit in the bearings, so they are driven into the front plate with a rubber mallet.

Then a pair of spacer sleeves are installed, and the rotors are pulled up tight against the bearings with an air wrench (a shoulder on the rotor shaft seats against the bearing race).

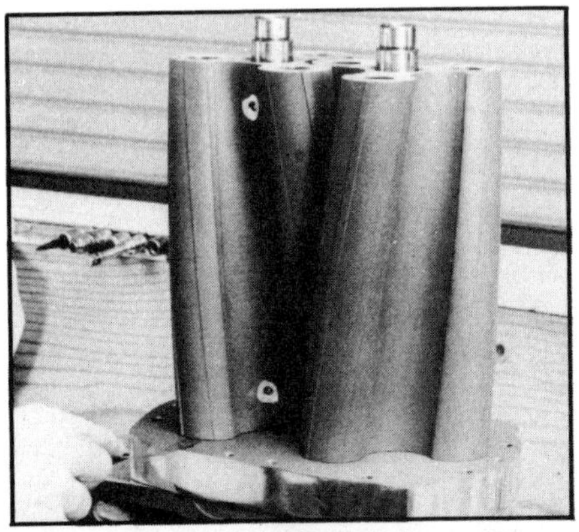

Using a feeler gauge, the front rotor-to-end plate clearance is measured. Mooneyham's front plate spaces the bearings out about .003-inch, which "tighten up" this clearance. On a street blower, Gene likes to set the right (drive) rotor at .007-inch and the left (driven) rotor at .008-inch to compensate for the tendency of the rotors to separate (one to move forward and one to move back) due to the helical cut of the drive gears.

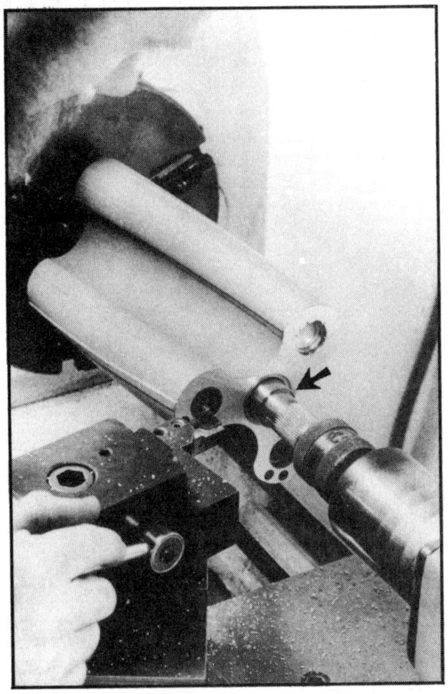

After initial measurement of the front rotor-to-end plate clearance, which should be less than the amount needed, Gene trims the end of the rotor in a lathe to get the clearance he wants. That is, if the clearance between the right rotor and the end plate measures .002-inch, he cuts .005-inch off the rotor. If a stock end plate is used, and the rotor-to-plate gap is too large, the stepped shoulder on the rotor shaft (arrow) would have to be turned down.

Once the front clearance is set, the case is mounted on a dummy rear plate (without bearings), and the rotors are set in the case to compare the length of the rotors to the length of the case.

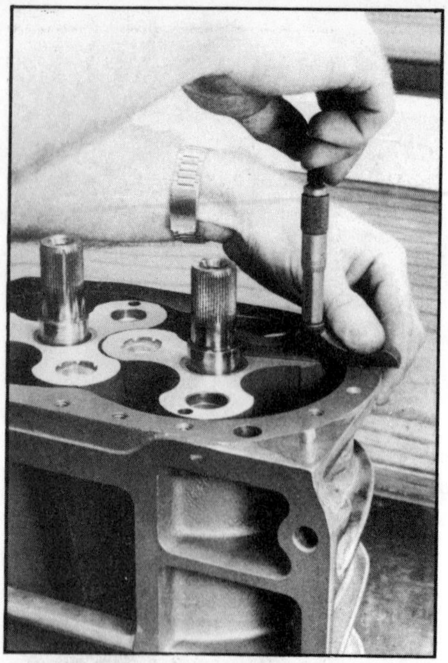

Using a depth micrometer, Gene measures the clearance between the top of the case and the end of the rotor. Remember that the case was trimmed about .010-inch already, so that the backs of the rotors can now be trimmed to set rotor-to-end plate clearance at the rear. Gene wants to get a .023-.024-inch measurement here, which will give .016-inch clearance at the back.

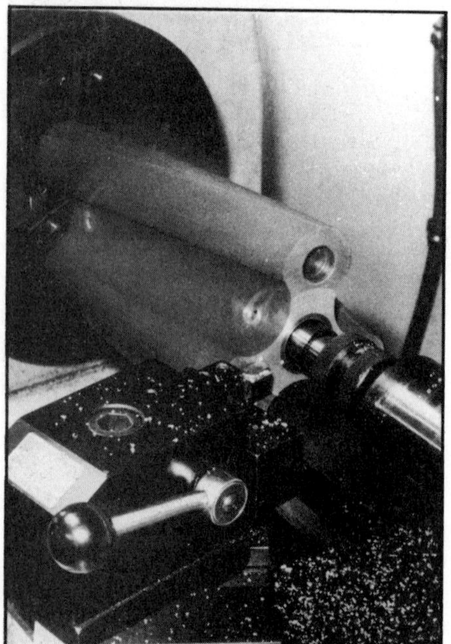

Then the rear of each rotor is trimmed in the lathe to give the proper clearance. Since the rotor is held in place, fore and aft, by the front bearing and end plate, the difference between rotor length and case length determines the rear rotor-to-end plate clearance. If for some reason the measurement in the prior step were too large (that is, if the rotor is too short), the end of the case would have to be milled instead.

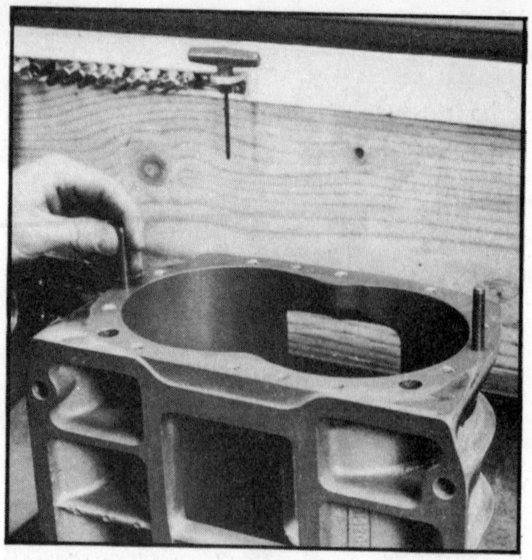

After machining of the rotors, the blower is assembled to check all clearances. First a set of stock dowel pins is inserted in the rear of the case...

...And the rear bearing plate, with bearings, is installed and snugged up with a couple of bolts.

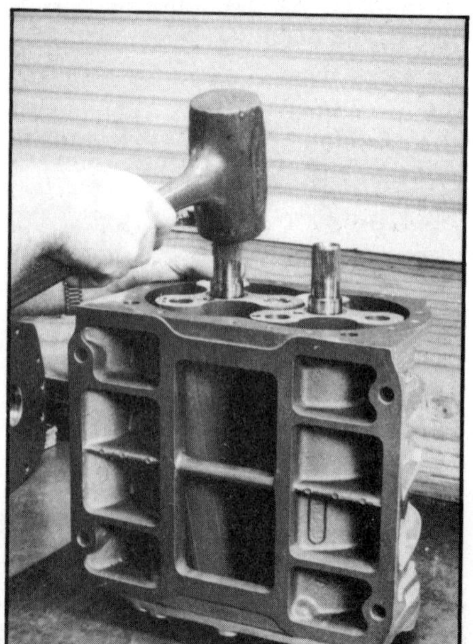

The blower is inverted and the rotors are tapped into the rear bearings.

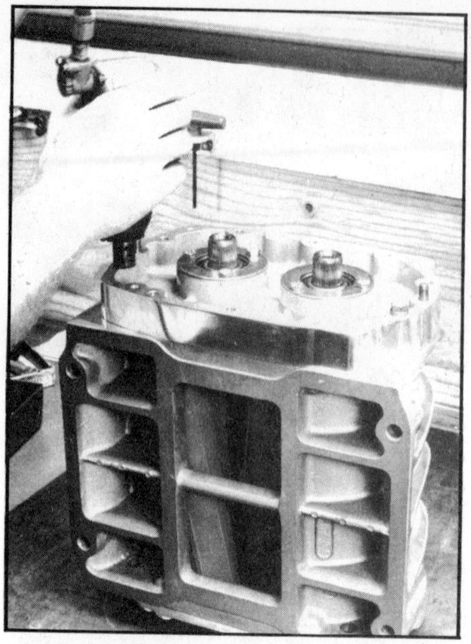

Then the front plate is installed with stock dowels and a couple of bolts. Once the end plates are tight, the dowels are removed.

The rotors are pulled up against the bearings with an air wrench.

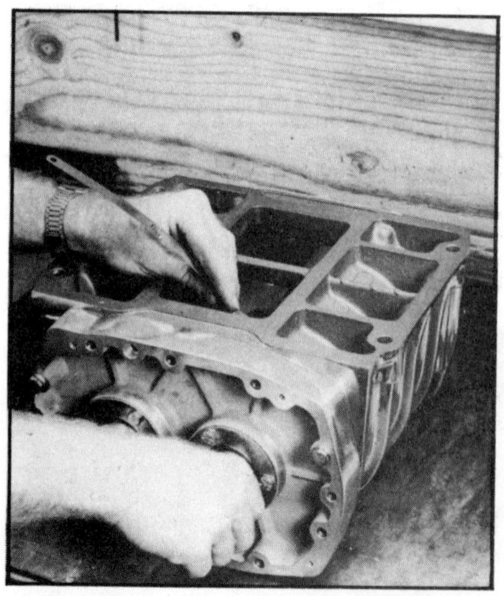

And then Gene uses a feeler gauge to determine the rotor-to-case clearance at each corner of the blower outlet.

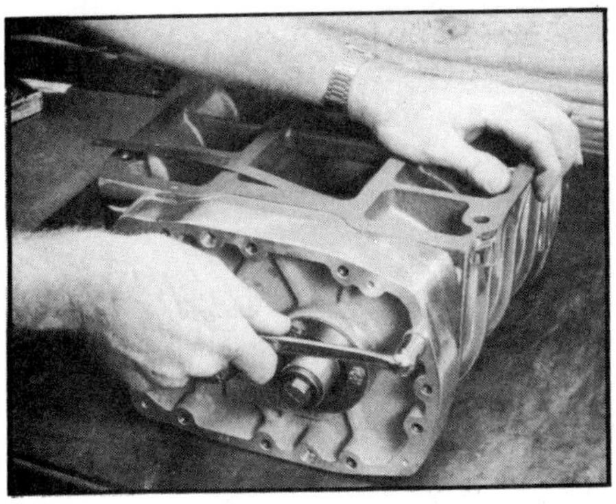

With the dowel pins removed from the end plates, the bolts can be loosened slightly...

...And the end plates can be tapped gently up or down to position the rotors in the case until the clearance is correct. The bottom rotor-to-case clearance is critical, since this is the side which must seal against manifold pressure. Gene sets it at .004-inch for street on rotors without Teflon. The top (inlet) rotor-to-case clearance is obviously determined by the bottom setting (decreasing the bottom clearance increases the top clearance), but the clearance on the inlet side is not very important—besides, the only way to change it, once the bottom clearance is set, is to make the rotors bigger or the case smaller...neither of which is feasible.

Once the end plates have been positioned to set the bottom clearance, Gene uses a long feeler gauge to check rotor-to-case clearance at each side of the case (inside the case at the location where he is pointing). The actual measurement isn't important here; the idea is to make the clearance the same on each side to center the rotors in the case.

So the end plates must be tapped right or left until the side clearance is equal.

But this always upsets the bottom clearance, so this must be re-checked and reset, and the process repeated until both are correct at the same time. When they are, the end plates can be tightened up with a few bolts to hold them firmly in place.

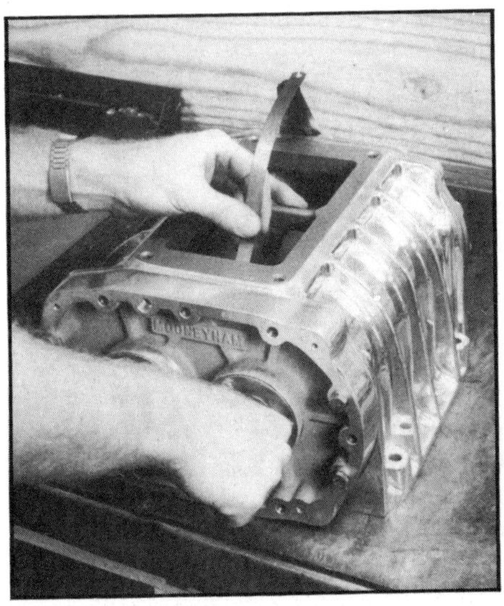

Then Gene re-checks the front rotor-to-end plate clearance with the blower assembled, just to make sure his prior calculations and machining were correct.

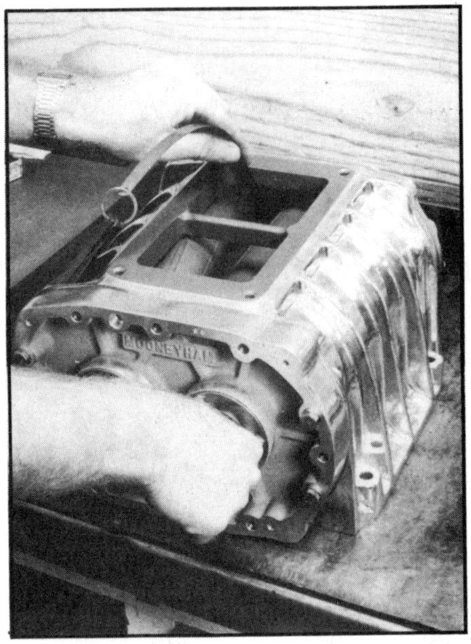

And does the same for the rear rotor-to-end plate gap. Hopefully they're correct; if not, the blower has to come all apart again and the last several steps repeated.

With the other clearances set, the rotor drive gears are installed and bolted in place. The meshing of the teeth on the gears "time" the rotors in relation to each other—that is, they determine how close the rotor lobes will mesh as they rotate.

Turning the rotors backwards, Gene inserts a long feeler gauge between the rotors to measure the clearance "up the middle." The interlocking of the rotor lobes in the middle keeps manifold pressure from leaking back up through the blower. Unfortunately the total clearance here (above and below each rotor edge) cannot be tightened up without making the rotors bigger all around (which anodizing helps to do, slightly). The main purpose for setting this clearance in an automotive blower is to time the rotors so that they don't hit each other.

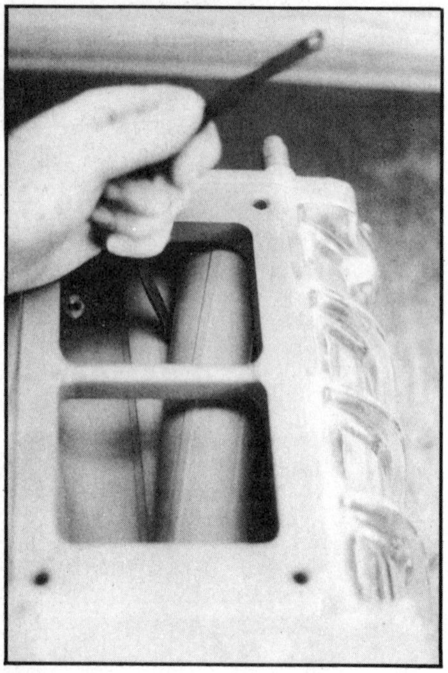

This close-up shows how the rotor-to-rotor measurement is taken. The feeler gauge (it must be pliable or properly bent) is inserted between the top edge of a left rotor lobe, and the bottom edge of the mating right rotor lobe as the rotors are turned backwards over each other. Then the left rotor is spun around so that the clearance between the bottom edge of the same lobe and the mating edge of the right rotor can be checked. That is, the clearance is checked at the leading and trailing edge of each lobe on each rotor (these are known as the C and CC clearances).

If the C and CC clearances are different—that is, for instance, the clearance at the leading edges of the left rotor lobes is .005-inch and the clearance at the trailing edges of the left rotor lobes is .025-inch, you want to advance the timing of the right rotor slightly to equalize the clearance on either side of all lobes. If this sounds confusing, it's just one reason why you should let a pro build your blower. Here Gene is using special pullers to remove the gears from the splined rotor shafts. Since tolerances vary in stock GM gears, Gene often tries several different gears to get a pair which "time" properly.

Otherwise, the way to alter rotor timing is to cut some of the shoulder off of the back side of the gear. Since the splines on the shaft are straight, and the gear teeth are helical, moving the gear in slightly on the shaft will advance or retard one rotor in relation to the other. For instance, trimming the back of the right gear (as you're looking at the blower) will turn the right rotor clockwise. Diesel mechanics put shims behind the gears, instead, to alter rotor timing the opposite direction. This is a simpler operation, but it is not recommended for performance blowers since the shims tend to get chewed up between the gears and the bearings under hard use.

Once all of the clearance-setting is complete, and the end plates are bolted just where he wants them, Gene drills through the dowel holes in the plate and the case with a larger bit to align the holes.

Next he reams the holes in both end plates and the case for a .0005-inch press fit for an oversize 3/8-inch dowel pin.

Then the entire blower (except pressed-in bearings) has to come apart once again.

And Gene reams the pin holes in the case to 3/8-inch so that the pins will slip fit in the case, but will stay pressed in the end plates. Thus the blower can be taken apart and re-assembled without changing the clearances that have been set.

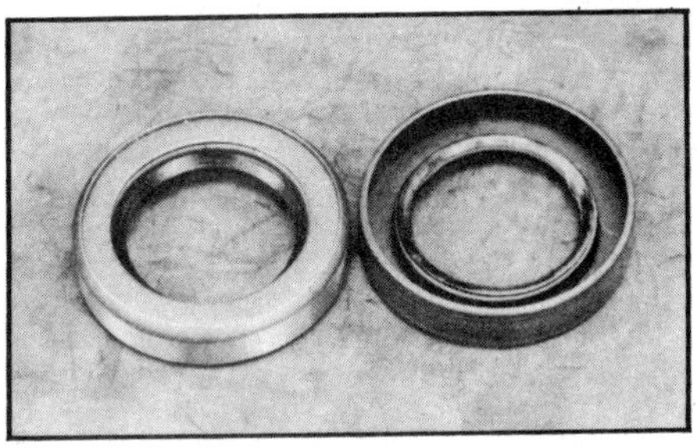

Good shaft seals are very important in a street blower. Gene uses a good double-wall neoprene seal with a stronger-than-stock spring. The stock GMC seal, on the right, has the lip vulcanized to the housing, and it tends to come apart under high speed automotive use.

After hand-packing the rear bearings with wheel-bearing grease, Gene presses the shaft seals into the front and rear bearing plates, with the lip facing the bearing so that oil is kept out of the blower case.

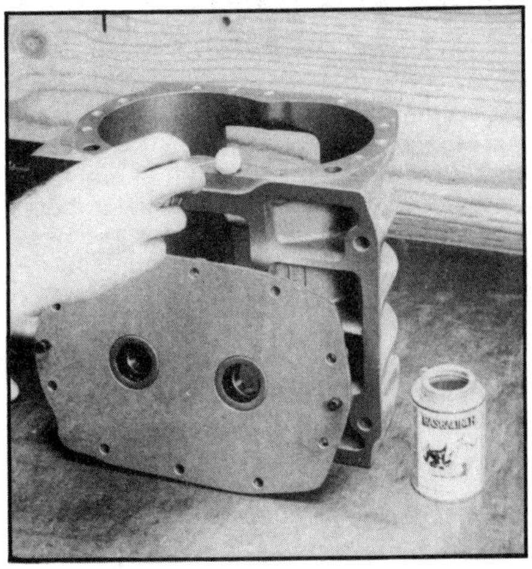

The blower assembles without gaskets, but Gene uses a thin layer of gasket sealer between the case and end plates.

The end plate is finally cinched up with a fresh set of Allen bolts.

The rotors are dropped into the case for the last time and tapped into the rear bearings.

Little Gene uses his big rubber hammer to tap the front bearing plate down over the rotor shafts before bolting it in place.

Finally the stock, polished front cover is bolted in place, along with a Mooneyham triangle cover at the front and a pair of grease covers at the rear, and this good-looking 4-71 is ready for a drive kit, intake manifold, and carb mount to install it on your favorite street machine.

HOW TO CHECK A USED BLOWER

These days you are likely to see quite a few used GMC superchargers offered for sale at swap meets, at speed shops, and in classified ads. In some cases a used blower—especially one that comes complete with a good drive, manifold, carbs, linkage, and so on to fit on your engine—can be a good bargain. But more times than not, a used blower is being sold because it is "used up." The problem is that you often cannot tell, just by looking, whether a used GMC is worth the asking price.

There are three categories of used GMC's, and this includes street or race prepped superchargers as well as stockers straight off of a diesel. The first category are blowers which are good enough to bolt on a street motor and run as-is. Very few used blowers fall into this group, and it often takes the feel and the eye of a blower expert to determine that a used blower is good. The second category are blowers which can be rebuilt for good street use. When shopping for a used blower, your best bet is to assume that you will have to have the blower "freshened up"—and pay for it accordingly. If the blower has already been set up for street use, you might get away with having just the bearings and seals replaced and having the clearances rechecked. But if the blower shop has to reset clearances, they will have to go through all the steps outlined in the previous section. In essence, a good rebuildable GMC is basically a good "core," that is, a good case, rotors, gears, front cover, and possibly end plates. In this case, a good stock diesel blower is just as good as one that came off a hot rod...perhaps better.

The third category of used blowers is for those which are not rebuildable. These are pure junk and are good for nothing other than wall decorations or door stops. The case and the rotors are the essential ingredients of any GMC supercharger—the other components are fairly inexpensive to replace. Neither rotors nor cases can be repaired; and if one is damaged, usually the whole blower is no good. Hardly anybody buys new rotors or cases from GMC because they are so expensive. So the plan is to find a good core at a reasonable price and then have it gone through by a reputable blower shop.

As we said, it really takes a trained hand and eye to determine whether a used blower is rebuildable or not, but there are some sure signs that a blower is *not* worth buying. In the following photos Don Hampton shows several ways to check out a used blower, many of which you can do on the spot (i.e., at the swap meet). However, a really good blower inspection requires disassembly and cleaning of the parts. If possible, you would be smartest to buy any used blower only on the contingency that you be allowed to have a professional check it out and proclaim it healthy before you pay for it in full.

Buying a used blower at a swap meet can be a risky business. This 6-71 with dual carbs and manifold might be a great deal if the price is low enough, but you can't tell without checking it...or having an expert check it. Note that this blower has stock end plates and front cover, and few drive components.

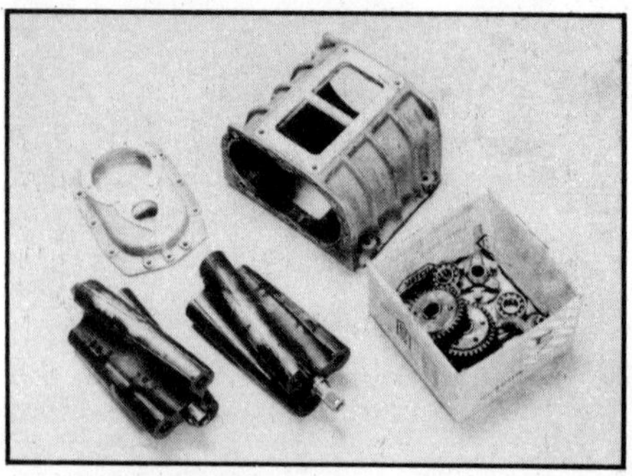

If you're going to have the blower rebuilt (which is the smartest approach), you might as well buy a good, stock, used core off of a diesel, like this 4-71, rather than paying for a supposedly "prepped" performance blower. However, note the oil on these rotors and the inside of the case.

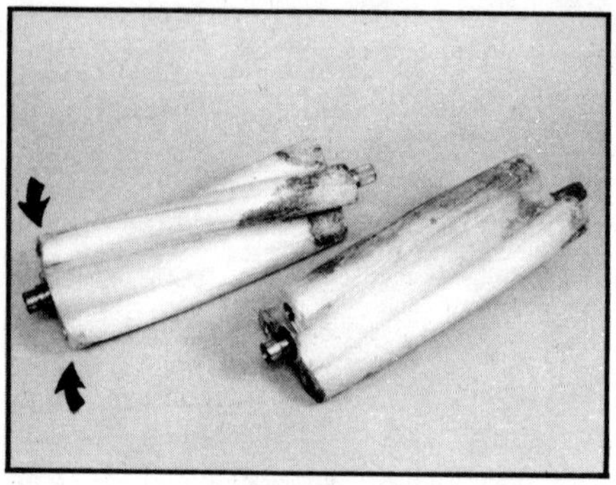

Once you clean the grime off of diesel rotors, you might find surprises like this. This 6-71 blower apparently sat outside for some time, and collected water in it, which corroded the aluminum rotors at one end. And at the other end, some very bad bearings allowed the tips of the lobes to be severely worn away (arrows). These rotors are complete junk.

As an example of a typical used blower, Don Hampton picked this old Danekas race 6-71. Begin with a complete visual inspection of the case, end plates, and rotors looking for cracks, obvious warpage, or stripped bolt holes. This one appears fine...but appearances can be deceiving.

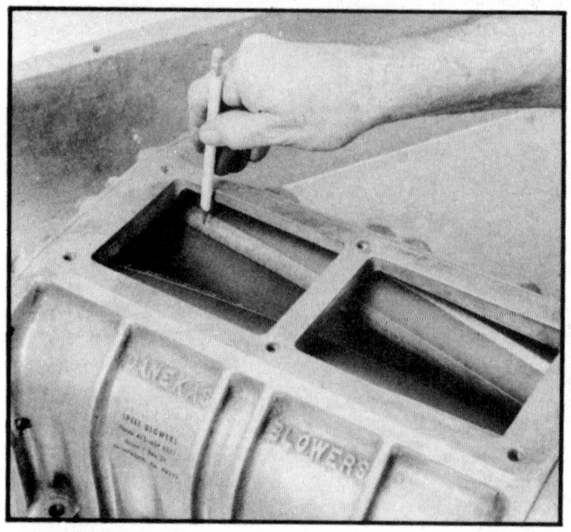

Hampton suggests you be very wary of used race blowers, since they are constantly subjected to severe operating conditions, and most aren't put up for sale until they are no longer rebuildable. One of the first places to check for damage on any blower is the sealing edge of the rotors, as shown. These are gouged and worn.

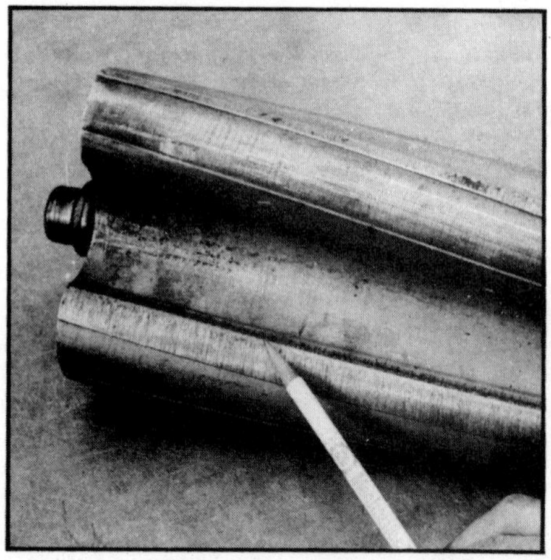

Here Don uses a loose rotor to show what you *don't* want to see. Pay particular attention to the edge closest to the "valley" on either side of each lobe—this edge should not be rounded off.

Most people first look at the outer edge of the rotor lobe, but this is usually not where wear is first noticeable (the rotors will usually hit each other before they rub on the case wall). The outer lip, or flat raised edge, of the rotor should be at least .080 to .090-inch tall and should not have any major gouges or dings in it. This one is passable.

Backfires are the nemesis of any blower, but especially one from a fueler. A backfire will force the rotors up in the case, causing them to hit the case wall along the top. Feel with your hand along the inner wall near the inlet opening, as shown, to feel for gouges or wear caused by a backfire. This blower failed this test. In fact, its rotors have been severely warped by a backfire sometime in the past.

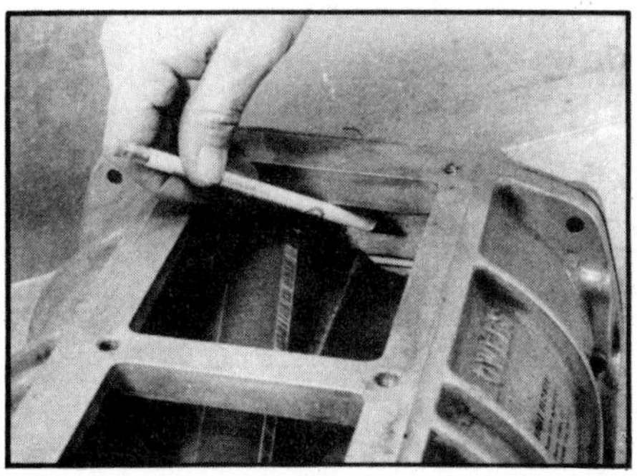

Visually inspect both end plates for wear or deep scratches.

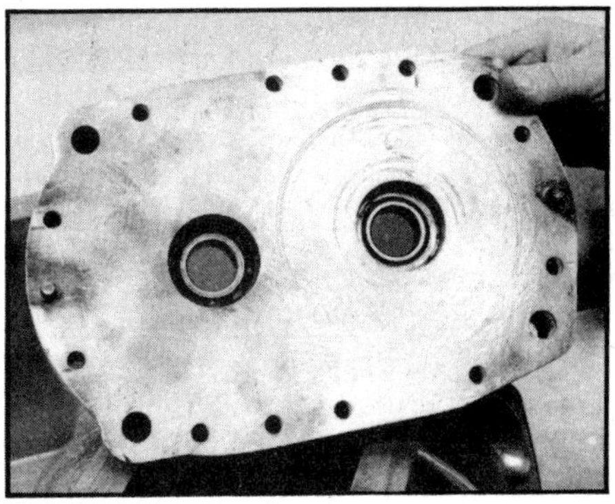

If an end plate is gouged like this one, the blower is not fit for use as-is. However, both the end plate and the end of the rotor (which would also be worn) could be machined smooth again when the blower is rebuilt, so this is not a serious problem...provided the rest of the rotors, and the case, are not worn.

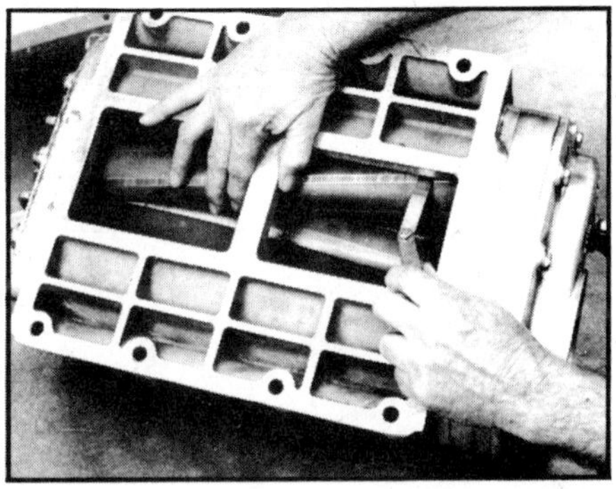

Finally, bring along a set of feeler gauges to check a few clearances. Your purpose is not to see whether the blower clearances are adjusted properly, but to see whether the total clearances around the rotors are within acceptable limits. If not, the rotors, the case, or both are worn beyond repair. Start by checking the rotor-to-case clearance at the outlet (bottom) edges. Measure at all four corners of the outlet opening. Clearance here should be in the neighborhood of .006-.008-inch. If you measure .010-inch clearance at the bottom, and .030-inch or more at the top (along the inlet edges), the rotors and/or case are worn out.

Clearance between the rotors and the end plate should be .018 to .020-inch maximum at the rear, .008 to .010-inch at the front. However, these measurements are not critical if the blower is to be rebuilt.

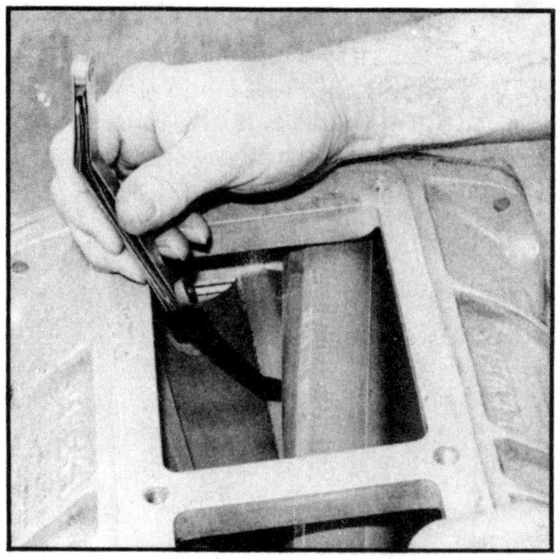

Finally, check the rotor-to-rotor (C and CC) clearance up the middle. This is a difficult measurement to make if you aren't used to the "feel" of the fit of the gauge between the rotors. You want to measure the clearance at the leading and the trailing edge of each rotor lobe, which should be approximately .020-inch each, or a total of .040-inch. If the total of both the C and CC clearances for any lobe is more than .060-inch, the blower is worn out and not worth buying.

Hampton bends the ends of his feeler gauges as shown to fit the curvature of the rotor lobes in order to get a proper reading between rotors.

On this open blower, Don shows where the clearance between rotors is measured, with the rotors positioned as shown. Then the left rotor would have to be rotated all the way around to measure the other side of the same lobe (arrow), and the feeler gauge would have to be inserted from the right side. If it sounds confusing (and it can be), see if you can take the blower to a shop to have it checked before you put your cash on the table.

PART 2: THE BLOWER BUILDERS

At the time this book is being written, interest in street supercharging is at an all-time high. Five to ten years ago, when street rodders were just beginning to experiment freely with GMC blowers, and to accept them as both practical and attention-getting, there were very few blower products to meet the rapidly growing demand. A handful of well-respected GMC blower rebuilders had been setting up superchargers for race cars for several years, and they could be counted on to turn out a good Jimmy for the street, but *adapting* the blower to a street motor was another matter. Other than the old Cragar V-belt drive for 4-71's, none of the blower manifolds or drive assemblies on the market made allowances for street necessities like a water pump, coolant plumbing into the block, a thermostat, an alternator, or a fan. Nor was there much available for mounting carburetors, rather than injectors, atop the blower. And if you wanted to bolt a Jimmy onto anything other than a small block Chevy, a Chrysler hemi, or perhaps an old Oldsmobile, forget it...nobody made any sort of manifold or drive, for street or for racing.

So street rodders learned how to adapt what they had. Cragar and Weiand 4-71 manifolds for small block Chevies were popular since they incorporated a standard water inlet and thermostat housing, and afforded room for a stock distributor. For other engines, rodders drilled and tapped holes directly into the heads for water flow to the radiator. Drive assemblies could be adapted from race parts on hand, and brackets could be handmade for engine accessories.

But it wasn't long before a few enterprising machinists saw the demand for street blower drive assemblies, and began making custom drive kits to adapt a customer's GMC blower, on an available manifold, to a street motor (usually a small block Chevy). Within a few years a couple of these machinists began devising methods for modifying existing aftermarket single or dual four-barrel intake manifolds to accept bolt-on plates to mount 4-71 or 6-71 GMC's, thus allowing blowers to be mounted on a much larger variety of street motors.

Now several of these manufacturers are finally offering complete bolt-on blower kits, consisting of a manifold, drive, a rebuilt blower, carb adapter plate, linkage, and everything else you might need other than carburetors (in most cases). Not only that, but suddenly we are seeing a host of new superchargers being introduced on the market, as well as the return of a few venerable ones—all of which are intended primarily for street applications. Is it surprising that everybody is getting the same message?

Supercharger Engineering is the only company currently offering the plentiful, slightly-smaller V6-71 GMC blower, complete with a drive kit (partially shown here) and low profile manifold for small block Chevies.

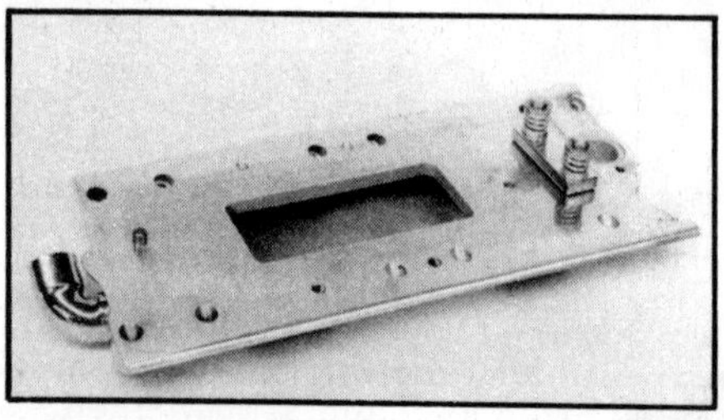

Dyer's new low profile manifold for his V6-71 and DMS-1 blowers is cast of aluminum and drops the blower down between the heads for increased hood clearance.

SUPERCHARGER ENGINEERING

Formerly known as Dyer's Machine Service, Gary Dyer's blower emporium located on the southern outskirts of Chicago was the first in recent history to offer GMC blower manifolds and drive kits for virtually any current V-8 engine. Dyer's "Street Charger" kits, introduced in the mid-1970's, certainly played a major role in popularizing the GMC blower for street machines, especially for cars with 440 Mopars, 351 Clevelands, small block Fords, Oldsmobiles, Pontiacs, AMC's, and other "offbrand" engines.

To make blower manifolds for a large variety of engines—manifolds incorporating water necks and other provisions for street use—without incurring exceedingly expensive tooling costs for casting each manifold separately, Dyer pioneered the modification of existing aftermarket intake manifolds to accept GMC blowers. Starting with a single or dual four-barrel aluminum manifold, Gary would machine the top flat, slightly enlarge the carburetor opening, and then bolt a flat aluminum plate onto the manifold to accept the blower.

Such a blower manifold, especially for a 6-71 Jimmy on a single four-barrel manifold base, may not be the optimum design (although Dyer claims the single four-barrel intakes will flow 2200 cfm, with the dual 4-bbls handling double that), but at least these modified intakes finally allowed a practical way to put street blowers on several popular makes of engines for which no other blower equipment was available. And for street set-ups, they certainly work well enough.

Dyer still hand fabricates several of his blower intakes, either on single 4-bbl, dual 4-bbl, or even tunnel ram manifold bases. However, he now uses a universal 6-71 mounting base casting which can be welded onto liberally opened-up manifolds to give a strong, one-piece, good-flowing blower intake for a variety of motors.

The vast majority of Supercharger Engineering's kits are for 6-71 GMC's, other than a 4-71 kit for the small block Chevy which uses a Weiand street blower intake manifold. However, Dyer has also recently pioneered the use of the V6-71 GMC blower for street engines. This blower is used by GM on V6 diesels (rather than inlines), and is very similar in design to the 6-71 except that it is slightly shorter and can be quickly identified by a different rib pat-

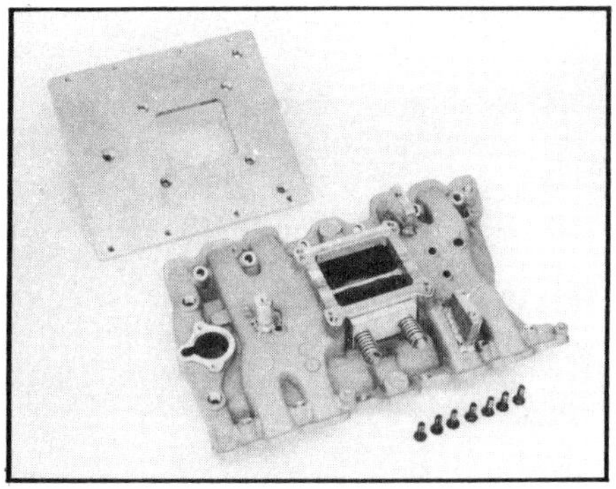

The ingenious idea that launched Supercharger Engineering was to modify existing 4-barrel intake manifolds to accept a flat machined plate onto which a 6-71 Jimmy could be mounted. Thus Dyer could offer blowers for Oldsmobiles, Mopars, Buicks, Pontiacs—whatever. Note support bosses welded to manifold to keep plate aligned, plus pop-off valve added to manifold.

Dyer still hand-fabricates blower manifolds for several engines, such as the Cleveland and small block Ford units shown, from single four, dual four, or even tunnel ram manifolds.

Recently, however, Dyer has had aluminum blower mounting plates cast, which can then be welded to a liberally cut-out dual four barrel intake manifold, as shown.

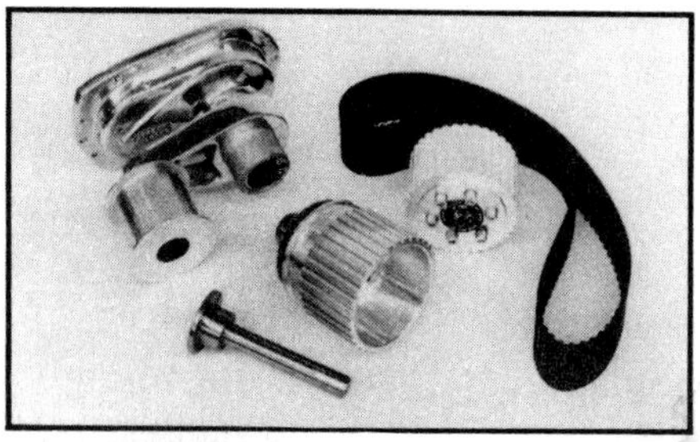

This is a typical Dyer GMC drive kit with a 3-inch wide Gilmer belt. Dyer makes his own blower front covers, which have an integral drive snout on which the idler pulley arm mounts.

tern on the side of the case, and angled attachment bolts which fasten near the center of the case. Dyer recommends the 6-71 for engines of 430 cubic inches or more, and the V6-71 for engines of less than 430 inches. Along with the installation and drive kits, Supercharger Engineering sells rebuilt and clearanced 4-71, 6-71, and V6-1 GMC's, fitted with their own front and rear covers.

Another recent development from Supercharger Engineering is a low profile cast aluminum manifold to mount the V6-71 blower on small block Chevies. With this intake, the blower sits 2-1/2 inches lower than on previous manifolds, allowing the complete package to fit under the hood of Chevy and GMC pickups, and possibly other earlier models of trucks and cars.

Finally, a project which has been in development at Dyer's for some time is a complete low profile blower package for small block Chevies and Buick V-6's. The major component of this package is a custom Roots-type blower designated the DMS-1 manufactured completely by Dyer's. The DMS-1 blower features custom-made 2-lobe rotors and a cast-in carburetor pad in the top of the case. The blower itself is only five inches tall, and the total height of the complete assembly, from manifold to carburetor top, is only 11-1/2 inches. The DMS-1 kit is designed as a low-profile, low-budget blower assembly which will install quickly and easily. The package includes a re-jetted, electric choke 625 cfm Carter AFB carburetor, a re-curved new distributor, and everything else necessary for a tailored, trouble-free assembly.

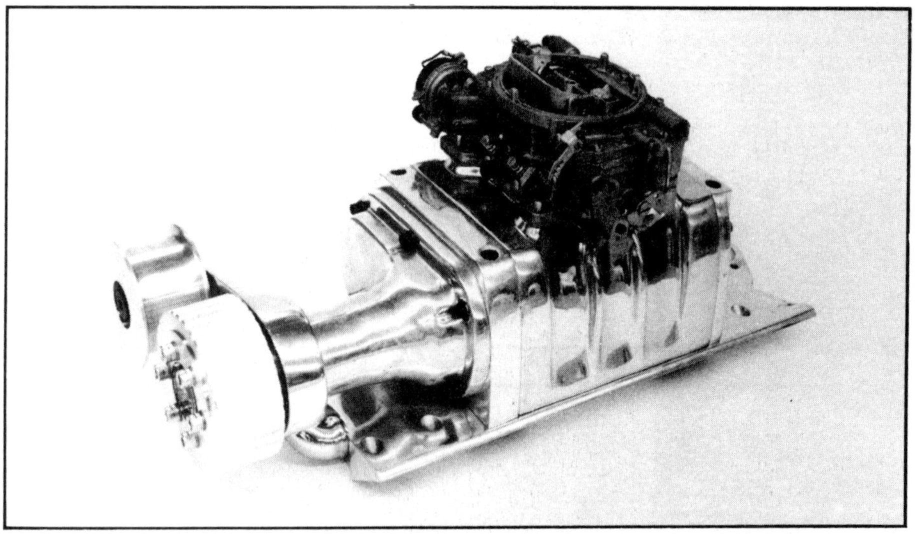

The latest offering from Supercharger Engineering is a completely new, small-displacement, in-house manufactured Roots type blower. It uses 2-lobe, straight-cut rotors, and comes complete with a low-profile manifold, drive kit, and reworked AFB 4-barrel.

Currently available in kits to fit small block Chevies and Buick V-6's, the DMS-1 blower was designed first of all to be a low-priced, complete, bolt-on package, and secondly to be an extremely low-profile assembly, as you can see on this Chevy.

A 6-71 with dual Holleys on a big block Mopar wedge is typical of BDS drive kits and accessories.

BDS uses a cast aluminum 6-71 blower base which they can weld to a variety of machined dual four barrel manifolds to allow blower installations on most any engine. Shown are 6-71 intakes for 455 Oldsmobiles and Pontiacs.

BDS also casts several complete blower manifolds, for engines ranging from big block Chevies to V-6's. This is a big block Chevy intake which will accommodate 6-71 to 10-71 blowers.

BLOWER DRIVE SERVICE

Other than Supercharger Engineering, most all of the major GMC blower rebuilders and drive manufacturers are located in California. Blower Drive Service, located in Whittier and operated by the collective Railsback family, is very probably the most complete blower shop anywhere. Like Dyer's, they specialize in fitting GMC blowers to a wide assortment of engines using modified manifolds to which they weld their own cast blower base. As of this writing they offered complete GMC blower kits, including a rebuilt blower, drive kit, manifold, and most accessories short of carbs and linkage, for big and small block Chevies, all modern big and small block Fords, big and small block Oldsmobiles, 326 to 455 Pontiacs, big and small block Buicks, big and small block Mopars, and all AMC V-8's.

Besides the V-8 line, BDS also makes complete 3-71 or 4-71 blower kits for Buick V-6's and 90-degree Chevy V-6's, as well as special kits using Magnacharger or Penco blowers for Ford/Capri V-6's, 60-degree Chevy V-6's, 2000cc Fords, and Datsun fours. In addition to rebuilding GMC blowers in-house, they also carry Magnachargers and other brands of blowers, and will custom fit blowers to other engines on a one-off basis.

The line of Blower Drive Service accessories is quite extensive. Besides their own 2- or 3-inch wide, 1/2-inch pitch drive assemblies, modified or cast intake manifolds, rebuilt blowers, or complete kits, BDS also sells blower cams and kits, blower pistons, carburetor linkage, fuel line kits, aluminum air scoops, and several other products. They also sell "blueprinted" Holley 600 to 850 cfm carburetors, and offer a carburetor rebuilding service for blower applications. Plus BDS is unique in offering blower fuel injection systems, in either downdraft or bug catcher styles, specially tailored for use on street-driven engines. See the section on carburetion for more details.

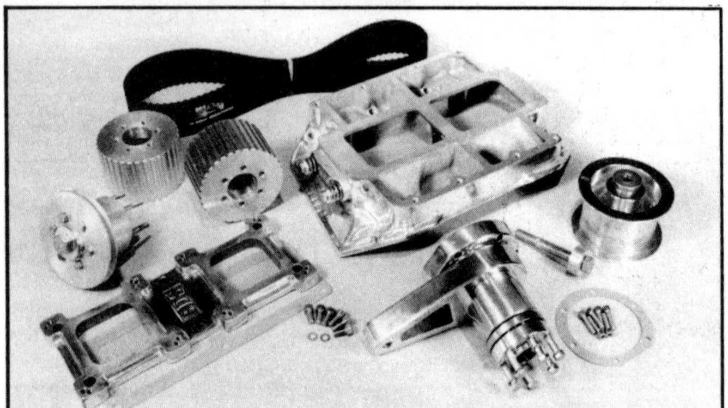

Blower Drive Service specializes in complete street blower kits. This set-up for a 440 Mopar includes a one-piece cast manifold, a 3-inch Gilmer drive, and an idler pulley bracket which mounts on the blower snout.

This BDS small block Chevy kit also uses a cast manifold, but mounts the idler pulley on a bracket (arrow) which mounts to the front of the blower.

WEIAND

Weiand Automotive of Los Angeles, in addition to making performance intake manifolds of all types, has made an excellent GMC blower drive assembly for many years. Just recently, however, they have added remanufactured and clearanced 6-71 GMC blowers to their product line, so that they can offer complete street blower packages for small block and big block Chevrolets consisting of blower, intake manifold, drive assembly, and carburetor adaptor. On a special order basis, Weiand will also sell complete 6-71 kits for 426 Chrysler hemi's and 4-71 kits for small block Chevrolets.

In terms of individual GMC blower components, Weiand offers 6-71 blower manifolds for big and small block Chevies, 426 Chrysler hemi's, and 354-392 Chryslers, plus a 4-71 manifold for Chevy smallblocks. The 4-71 manifold incorporates a standard thermostat-type water neck, but the 6-71 Chevy manifolds have two 3/4-inch water outlets at the front, requiring connection to a remote thermostat housing with two small hoses (included in the kit) for normal radiator filling, etc.

This is a complete street 6-71 kit from Weiand, including a rebuilt blower with a custom front cover and a cast intake manifold. Weiand's 6-71 manifolds do not incorporate street-type water necks, however, and require the use of a remote thermostat housing (arrow).

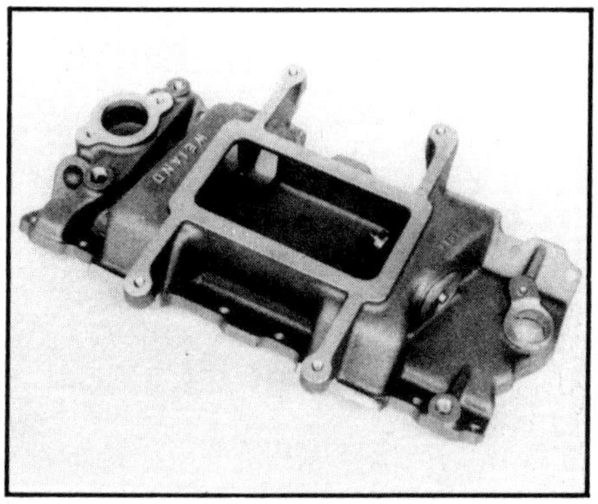

An excellent street manifold from Weiand is their 4-71 intake for small block Chevies. It has full street hookups.

Weiand makes two versions of their drive kit. Kit "A" is the street unit, which includes a cast blower front cover that mounts the idler pulley bracket, and a V-belt lower pulley that, along with the lower blower belt pulley, is designed to bolt to the stock harmonic balancer. Weiand's Kit "B" is the competition version, which has no provision for accessory belts, and which uses a short drive snout which mounts on the stock GMC front cover. Otherwise both drives are similar.

Other Weiand blower accessories include a heavy duty front bearing plate and a thin, finned rear cover for GMC blowers, and single or dual 4-bbl carb adapters for 4-71's and a dual 4-bbl adapter for 6-71's.

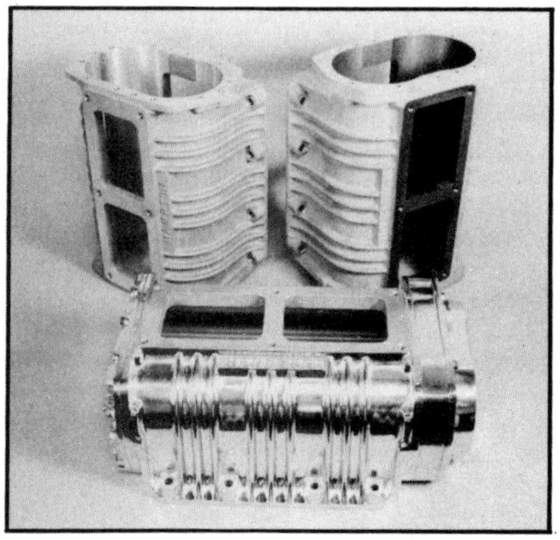

Hampton has made his own blower cases for years—the 9-rib style is certainly distinctive. Behind the polished 6-71 are a rough 6-71 (left) and an 8-71 (right).

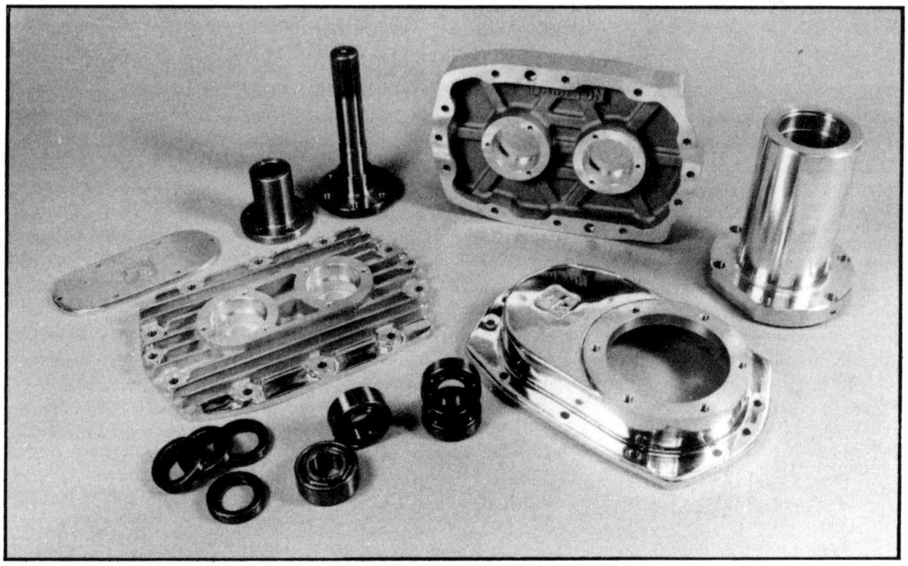

Don Hampton has probably the most complete line of custom-made GMC blower components. Besides the cases, he also manufactures his own front and rear bearing plates, front cover, drive shafts, snouts, bearings and seals (as shown), plus he is now making his own 2-lobe rotors for Jimmies.

DON HAMPTON

Don Hampton first gained drag racing recognition with a twin Chevy powered, supercharged, Fiat-bodied competition coupe in the early 60's, later converted to a funny car with a Corvette body. As you can imagine, he learned plenty about GMC blowers keeping two of them working properly at once. After retiring his twin Chevy, Hampton spent several years building GMC's for racers out of a well-equipped shop in his garage.

Just within the last couple of years, however, Hampton has opened a new shop in Downey, California, and has branched out into a serious street blower business, in addition to his continuing competition business. Previously Hampton just rebuilt blowers, and manufactured his own aluminum and magnesium cases, front bearing plates, end plates, and front covers. Now Don is making up manifolds, similar to those welded up by Supercharger Engineering and Blower Drive Service, to mount 4-71 or 6-71 GMC's on all sorts of popular V-8 engines.

In addition, Hampton also casts a variety of single or dual four-barrel carburetor adapters, and manufactures full street drive systems for numerous applications. Hampton also carries carburetor linkage kits, and manufactures a unique remote-mount, belt-drive water pump for blower installations which do not allow room for a stock water pump.

Like BDS and Dyer, Hampton also now casts his own aluminum weld-on 6-71 base (left), plus a 4-71 base which he uses for V-6's (right).

Hampton uses his weld-on base to adapt 6-71's to nearly any type of manifold. Shown in this sampling are a small block Ford and 351-Windsor.

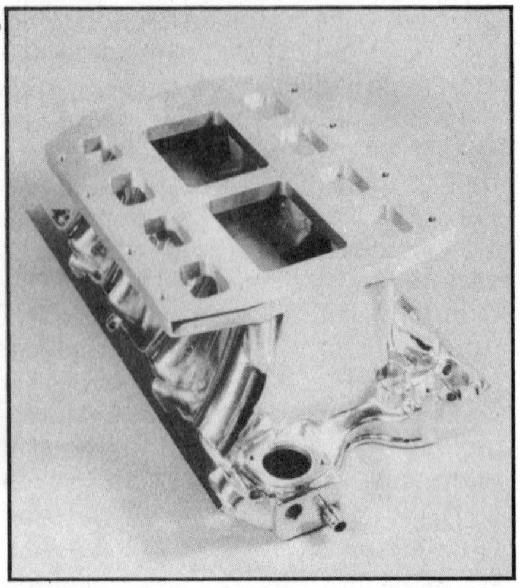

Don will also hand-fabricate blower applications, such as this 6-71 manifold for a 460 Ford which uses a milled flat aluminum plate welded to a tunnel ram base.

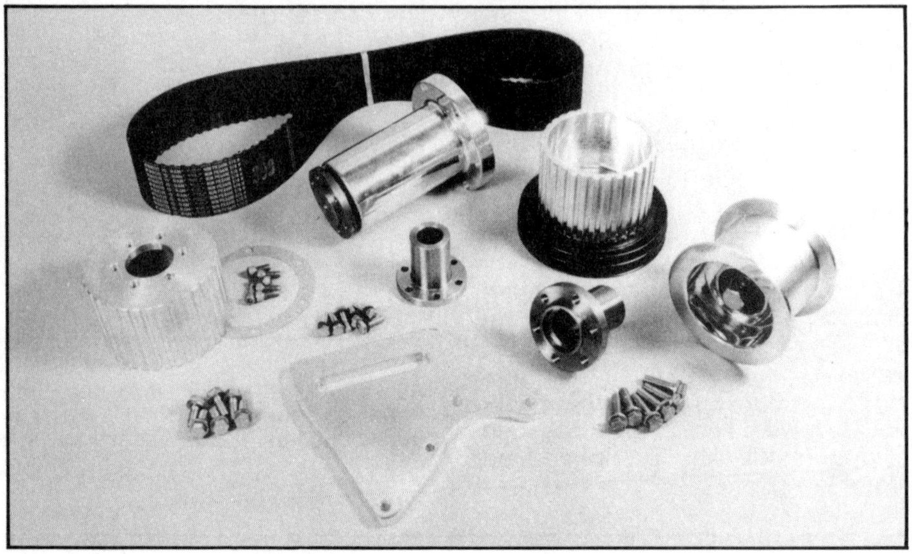

This is a typical Hampton 3-inch drive kit for a street set-up (this one is for a 351-C Ford). The idler pulley bracket mounts to the front of the block or water pump in most cases.

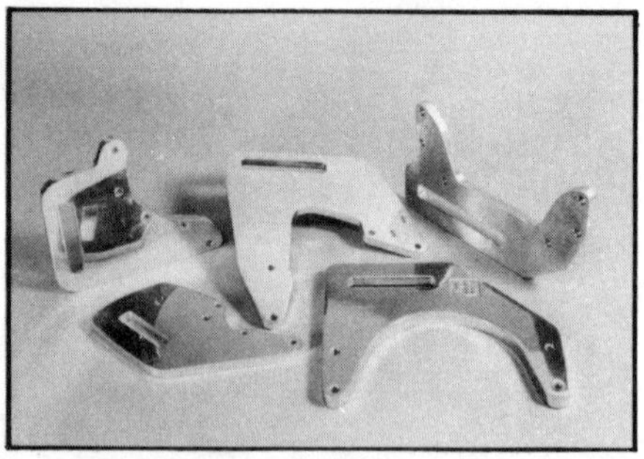

Hampton makes and stocks a wide variety of street drive accessories including stainless steel crank hubs and cast or machined idler brackets for numerous applications.

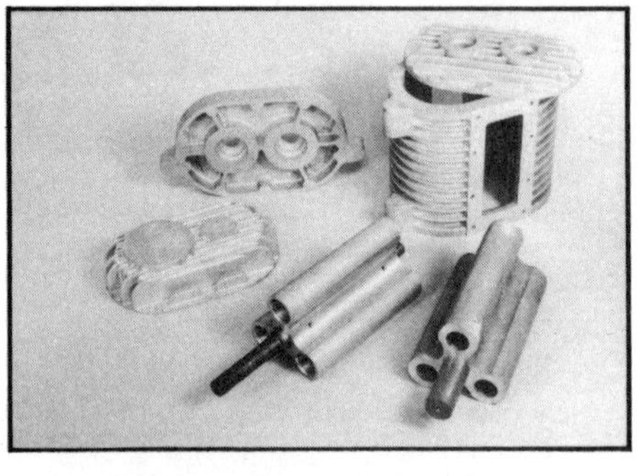

A project Don has waiting in the wings is this good-looking, 3-lobe, small displacement Roots type blower. But Don says he's so busy building GMC's for street machines, he hasn't had any time to progress with its development.

GENE MOONEYHAM

Through the 50's and early 60's this guy campaigned one of my all-time favorite drag cars, the blue, chopped, fenderless, blown-fuel-hemi-powered Mooneyham and Sharp '34 Ford. Today he is one of the most fastidious and highly respected of the GMC competition blower rebuilders.

Mooneyham's work is still primarily geared toward professional and amateur drag, boat, and tractor-pull competitors, and many of his current blowers are of the larger 8-71 through 12-71 sizes using his own aluminum cases. However, he is not averse to setting up stock-case GMC's for the street (as we show in the accompanying photo essay), and his prices for doing so are competitive with other's. Mooneyham, like most blower builders, is also used to shipping rebuilt blowers to all parts of the country.

Besides his own 6-71 through 12-71 aluminum cases, bearing plates, rear covers, and front covers, Mooneyham manufactures a popular big block Chevy intake manifold which is designed to reduce the tendency of blown big blocks to run lean at the forward cylinders. Mooneyham is also unique in manufacturing a water-cooled blower case, available in 6-71 or 8-71 sizes. This blower has a water jacket on either side of the case, which accepts cold water at two inlets near the top and filters it down the sides of the blower.

The water-cooled blower is is designed for use on boats, which have a ready source of cold water for cooling; it would not be practical without a very cold water source, and even then the amount of heat that could be absorbed from the inlet charge by coolant in the blower case is rather small. The problem is that most of the heating of the air occurs below the blower with a GMC, since the compression is being done primarily in the manifold.

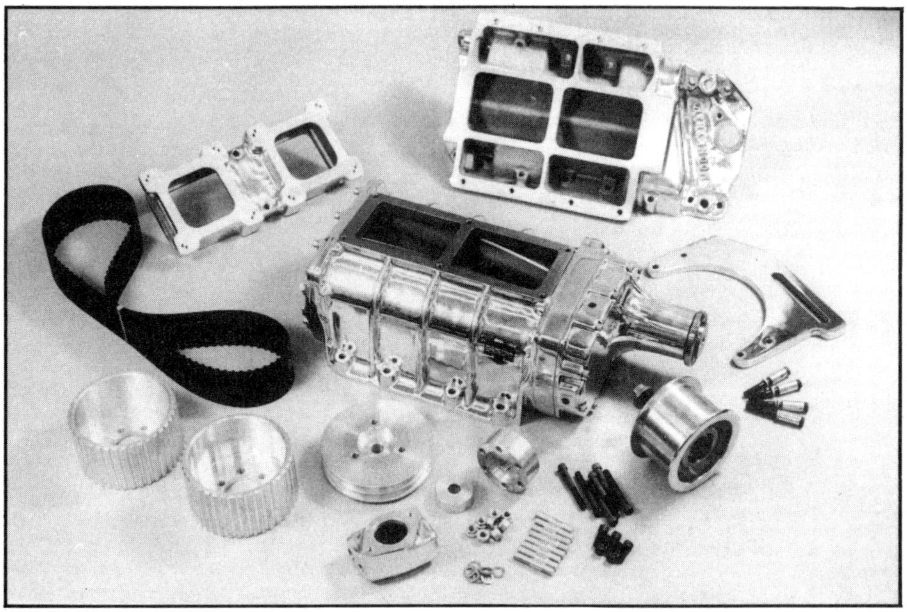

Although Mooneyham builds more competition blowers than street set-ups, he can offer complete packages for some engines. This is a stout, Teflon-lined 6-71 for a big block Chevy. Most of the drive kit, including the blower front cover, is from Weiand. However the manifold is a Mooneyham product.

Gene Mooneyham makes his own blower cases in several sizes. Shown here are a 10-71 (front) and a 12-71, which are strictly for racing. He also makes his own front and rear bearing plates and front covers.

A unique Mooneyham product is this water-cooled blower case for use on boats. Cold water is pumped in one orifice and out the other on each side to lower case temperature and help cool the induction charge.

BOWERS SUPERCHARGERS

Larry Bowers, whose shop is located in Chatsworth, California, on the north side of Los Angeles, specializes in race blowers and will build street units only as time permits. The proclaimed inventor of the 10-71 blower, Bowers manufactures his own cases in either magnesium or aluminum in 6-, 8-, 10-, or 12-71 sizes. He also makes his own front and rear covers, bearing plates, and drive assemblies—everything except manifolds.

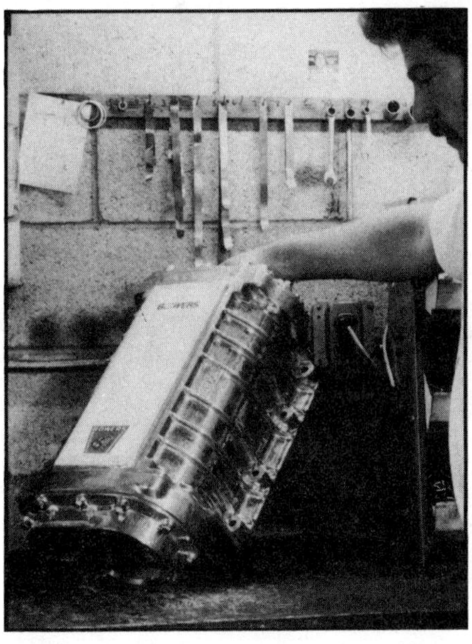

Larry Bowers makes several sizes of GMC-type blower cases in both aluminum and magnesium, as well as bearing plates, front covers, and other components.

LITTLEFIELD BLOWERS

Current Funny Car pilot Mert Littlefield also specializes primarily in competition GMC-type blowers. He manufactures his own distinctive blower cases in either aluminum or magnesium in all sizes from 6-71 through 14-71, as well as his own bearing plates, end plate, and front covers. For gasoline applications, Littlefield's major business is with ski boat owners, for which he builds 6-71's and 8-71's, primarily for big block Chevies. Mert will also build street blowers, using either his own cases or GMC's, but they are not his primary business and he does not offer any compatible street drive systems.

Littlefield does market Cragar competition blower drives and manifolds, however, plus he also manufactures his own drive pulleys in 8mm, 14mm, and 1/2-inch pitch to offer a wider assortment of drive ratios than those available from Cragar. In addition, Littlefield makes a line of dual four-barrel carburetor mounts of a tall, wide-spaced design intended primarily for boat motors.

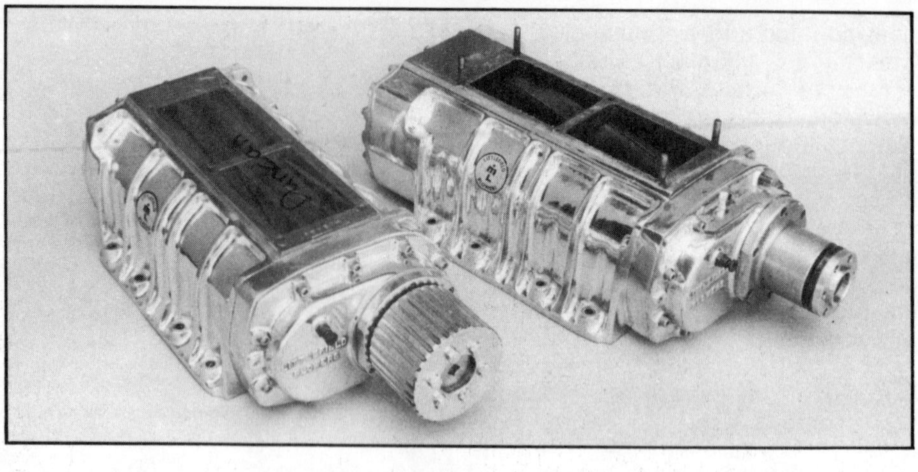

Mert builds lots of the new jumbo-size blowers for racers, especially alcohol dragsters and funny cars. This is a 10-71 and a giant 14-71. Littlefield uses Cragar drives and manifolds, however, for which he is a dealer.

Mert Littlefield is another of the major competition blower builders located in California. He builds his own cases, in either aluminum or magnesium, as well as bearing plates and front covers. This polished 6-71 is also fitted with one of Littlefield's wide-spread carb adapters which allows Holleys to be mounted in line.

This is an example of one of Mike Kuhl's "River Rat" blown gas Chevies, built for ski boats with one of his own 6-71 blowers. Note small cam-driven water pump and water fittings in cam cover and manifold.

MIKE KUHL

Mike Kuhl Racing Enterprises in Santa Ana, California, is primarily an engine building shop, but since they build so many supercharged drag racing and boat engines, they have become heavily involved in blower building as well. Recently Kuhl has acquired his own patterns for custom GMC-type blower cases. This unique design has rounded sides on the case and a completely flat, solid bottom (except for the blower outlet). These blowers are available in either aluminum or magnesium in a range of sizes. For drives, manifolds, and accessories Kuhl uses Cragar components, for which he is a distributor.

Although Mike Kuhl built his reputation on blown fuel and alcohol Chryslers and Donovans (which still account for a large portion of his business), he also specializes in blown gasoline big block Chevies—which he calls his "California River Rat"—for ski boats. For these engines Kuhl has designed an alternator drive, a neat cam-driven water pump, and other accessories which would also apply to street blown motors. For these engines Kuhl also makes a wide-spaced dual 4-barrel carburetor adapter and various styles of aluminum scoops. Of course, Kuhl builds complete street or boat supercharged engines, though primarily of the Chevy or Chrysler variety.

Close-up of installed Kuhl blower shows his wide-base carburetor manifold, Cragar drive, and Cragar blower manifold which has been tapped at the front for small water fittings (arrow).

The new Kuhl blower has a unique case and custom end plates. This one is made of polished and anodized magnesium and has a satin-like gold tinge.

Larson Engineering is a new source of rebuilt GMC blowers and custom-made drive systems located in the Midwest. Shown here is one of their hand-built drives, including a chrome blower belt shield, for a small block Chevy with a 6-71.

Larson is the company which originated the wide, 2-piece valve covers for Chevies (shown here), and they also offer several varieties of injector-like air cleaners to fit either single or dual 4-barrels.

LARSON ENGINEERING

Shortly before this book went to press we received word of another new GMC blower and installation kit manufacturer located in the Midwest. Larson Engineering of Terre Haute, Indiana—manufacturers of the recently-popular wide, two-piece, finned aluminum valve covers for small block Chevies—now offers a full line of Jimmy blower kits for big and small block Chevies, big and small block Fords, Buick V-8 and V-6, Oldsmobiles, Pontiacs, as well as other engines on a special order basis.

Larson blower kits are available with either 6-71 or 4-71 rebuilt and polished blowers, two or three inch drive belt and custom-machined pulleys, and single or dual quad carburetor mount. Larson also offers complete blower drive systems (less blower), individual drive components, "blueprinted" Holley 4-barrel carbs, and chrome blower belt shields. Of course they also offer their unique valve covers for big and small block Chevies and Buick V-6's, plus custom air cleaner assemblies for dual four-barrels.

The GMC blower rebuilders listed here are certainly not the only ones in the country who set up Jimmies for the street, but they are the only ones doing any major business in this field at the time of this book's publication. They are also the only GMC rebuilders who manufacture their own performance blower components—other than Ed Pink, Donovan, or a few others who deal strictly with competition superchargers. There are undoubtedly other drag blower rebuilders scattered around the country (or in foreign countries), and there may be a few who set-up GMC's for the street, such as Rudy Mortensen of Mortensen Engineering in San Carlos, California, but the ones discussed here are the primary street GMC rebuilders doing nation-wide business. Each of them regularly ships blowers and components to various parts of the United States, as well as foreign countries.

GMC BLOWER DRIVES AND ACCESSORIES

ISKENDERIAN

Isky Cams of Gardena, California, was one of the first companies to design and market Gilmer belt drive assemblies for GMC blowers back in the 1950's. Their current "forced induction" drive kits still reflect 1950's drag racing design, but they have modified their small block Chevy 2-inch wide belt drive kit to accommodate a stock water pump (using a large un-toothed aluminum pulley for the pump which is driven by the back side of the blower belt), and generator. This kit includes belt, pulleys, a very stout idler and bracket, crank hub with V-belt groove, and a blower drive hub, shaft, and snout. An Edelbrock 6-71 blower manifold is mandatory with this drive (they are available from Isky), and of course you must supply the blower, carburetors, carb adapter, and linkage. Isky also carries similar 2- or 3-inch drive assemblies for big block Chevies and for 1956-58 Chrysler hemi's, but these make no provision for a water pump or accessory drive.

The Isky drive uses the blower belt to turn the water pump, and this helps to keep everything close to the front of the engine, although you still need to make an alternator mount and drive. Isky also uses the Edelbrock intake manifold, which requires that the front of the heads be drilled and tapped for pipe fittings (arrow) to hook up to radiator top tank.

CRAGAR/MR. GASKET

For at least two decades the standard blower drive assembly in Top Fuel drag racing and most forms of all-out supercharged competition has been the Cragar set up. In fact, several of our current 6-71 street blower drive kits derive from the basic Cragar design. However, Cragar components are strictly of a competition nature, with no provisions whatsoever for street hookups such as radiator cooling, fan, water pump, or alternator drive.

Cragar competition kits include manifolds for 6-71 and 8-71 blowers to fit big and small block Chevies, 354-392 Chryslers, and 426 Chrysler hemi's; blower drive snout; belts and pulleys; and idler bracket. Cragar makes a wide assortment of 1/2-inch pitch and 8 mm blower pulleys and belts, and they are the only company which offers the larger 14 mm metric pulley and belt set-up (required by today's Top Fuel cars, but totally unnecessary on the street).

However, just as this book was going to press, Cragar Industries had been assumed by Mr. Gasket/American Industries of Cleveland, Ohio, and the future of the Cragar blower equipment was a bit uncertain. It is assumed that the Cragar competition blower manifolds and drive systems will remain available. However, Mr. Gasket has also announced that it will re-introduce the basic Cragar 6-71 competition blower drive in a modified form as the Cragar Street Drive, which will include provisions for a water pump, manifold water inlet, and alternator drive and mounting brackets.

Other blower-related products from Mr. Gasket include a variety of blower gaskets, dual four barrel carburetor linkage kits for either in-line or sideways actuation, and the new Mr. Gasket Street Scoop which bolts on top of dual four-barrels. Simulating a bug-catcher type blower injector, the Street Scoop has three butterflies which operate with the throttle linkage; it also comes with two paper-type filter elements which are concealed inside.

Mr. Gasket recently purchased Cragar, and will reportedly be offering new Cragar blower drive kits in street versions, as well as a full line of accessories, such as their popular Street Scoop.

Cragar also makes a wide assortment of Gilmer belts and pulleys. Shown, left to right, are 1/2-inch pitch, 8mm, and 14mm belts and pulleys. Cragar is the only company which makes the 14mm belts, but these are strictly for Top Fuelers. Street machines need nothing better than a 1/2-inch pitch drive, and most can get by with a 2-inch wide, rather than a 3-inch, belt.

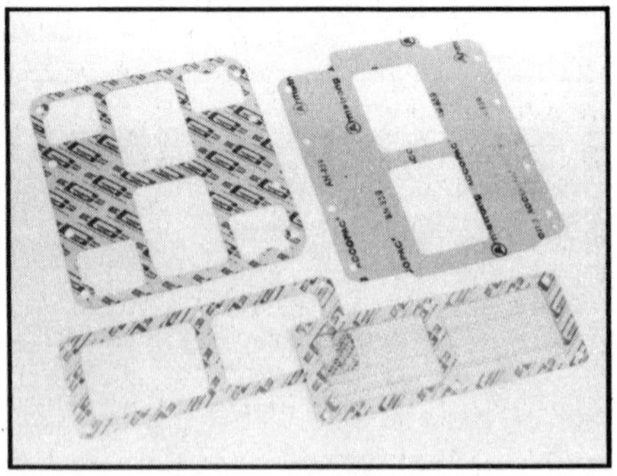

Of course Mr. Gasket offers a full line of blower gaskets for GMC blowers.

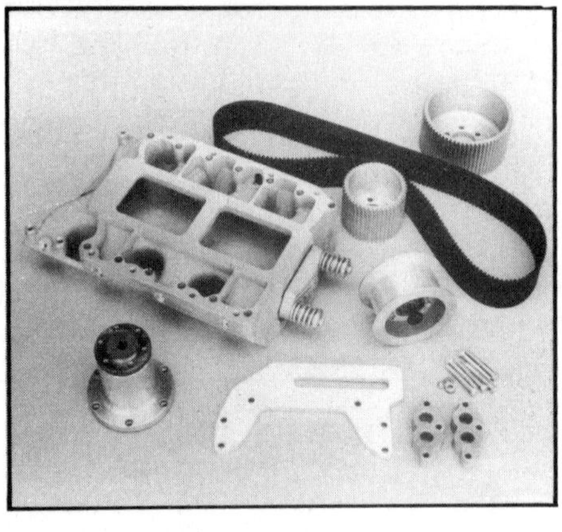

Cragar manifolds and drive systems have been the standard for racing applications for decades. Shown is a complete Cragar drive, with 8mm belt, for a Chrysler hemi. Many street systems have been patterned on the Cragar style, or may use some Cragar components, but Cragar has made no street systems themselves.

CUB BARNETT

The original Cragar 4-71 small block Chevy street blower set-up, consisting of a blower manifold, blower snout, and a triple V-belt blower drive system, was originally purchased by Andy Brizio and Cub Barnett when they owned Champion Speed Shop in South San Francisco, and this system was installed on several of Andy's "Instant T" kit car street rods in the 60's and early 70's.

This system is now being marketed by Cub Barnett, whose shop is located in Belmont, California. Cub also makes similar drive assemblies, using three V-belts to drive the blower and another for the water pump and alternator, for a 4-71 GMC on 289-302 small block Fords and for 6-71's on big block Chevies. The Ford set-up uses a Weiand hi-rise single four-barrel intake manifold, to which Cub bolts a flat aluminum blower-mounting plate and a pop-off valve. This design affords clearance for the front-mounted Ford distributor, and even allows distributor removal with the blower in place. The big block Chevy system uses a Weiand blower manifold. Barnett will supply complete supercharger systems for these engines, using blowers rebuilt by Rudy Mortensen.

EDELBROCK

Edelbrock still manufactures one blower manifold for the small block Chevy. It accepts a 6-71 GMC, and it has no provision for water outlet, requiring that the heads be drilled and tapped on the front for water hoses to the radiator.

Edelbrock has long offered a competition-style 6-71 intake for small block Chevies. It uses an internally-sprung intake valve as a pop-off valve, and it has no provision for radiator hook-up. It is also available through Isky Cams.

OFFENHAUSER

If you really want to put your 6-71 up in the air where everybody can see it, Offenhauser offers, on a custom-order basis, a 6-71 base which mounts atop their "Turbo-Thrust" tunnel ram manifold bottom section, giving a very tall, tunnel ram type blower manifold. This blower top can be added to manifold bases available for big block Chevrolets, "B series" Chryslers (383-440), 426 Chrysler Hemi's, and 429-460 Fords. One advantage of such a manifold is that it allows plenty of room for a front-mounted distributor. The disadvantage is that you would have to fabricate your own drive from existing components...and find a very long blower belt.

Offenhauser also lists a very large assortment of carburetor linkage, either in dual-quad kits or in individual pieces. Plus they make a cast aluminum remote water neck and thermostat housing to allow the use of a thermostat on street set-ups where the manifold has no such provision.

Although they are extremely uncommon in the U.S., Wade Roots-type superchargers are plentiful in England, where they are made. They come in two sizes, the larger RO34, which has a right-angle inlet housing, and the smaller RO20.

The unique characteristic of the Wade blower is that it uses straight-cut, 3-lobe rotors with diagonal inlet and outlet openings in the case (the outlet side is shown here). This gives the same result as the "twisted" rotors in a GMC, but makes machining of the rotors much easier.

PART 3: OTHER ROOTS-TYPE BLOWERS

THE WADE

In general we will not discuss European superchargers in this book, especially those of eras gone by. The Wade Roots-type blower actually falls into both of these categories, but I think it is worth a brief mention because (1) it was the most recent and most popular Roots blower used for automotive performance applications in England, and (2) it has an interesting engineering twist.

Wade blowers, like GMC's, were originally designed and used as scavenge pumps for two-stroke diesel engines. However, Wade converted two sizes of their blower, known as the RO34 and the RO20, to gasoline automotive engine use by substituting steel rotor support shafts for the aluminum shafts used in the diesel models. With the steel rotor shafts, these blowers reportedly need no reworking for automotive performance use if blower speed is kept below 5000 rpm.

The Wade blowers use a die cast aluminum case and extruded aluminum, three-lobe rotors. The longer RO34 has a displacement of 3400cc per revolution (207.5 cubic inches), and the RO20 displaces 2000cc (122 cubic inches). The unusual design characteristic is that, while the rotors are straight-cut (untwisted), both the intake and outlet openings in the blower case are cut on a diagonal, which ultimately gives the same result as putting a helical twist in the rotors, without incurring either the tremendous expense of machining helical rotors or the clearance problems such

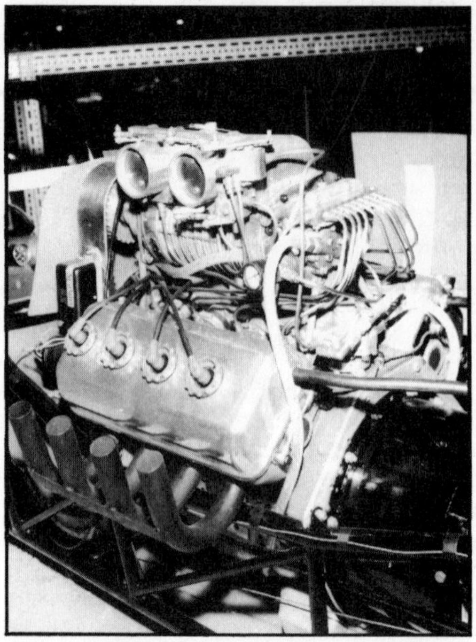

The English have come up with some interesting applications of the Wade over the years. Shown here are two injected, fuel-burning, mini-dragster installations, one on a 2.5-liter Daimler hemi, the other a front-mount on a little four cylinder.

rotors cause at high speeds and temperatures. The Wade's diagonal-cut ports, like the GMC's twisted rotors, reduce both the pulsing of airflow from the blower, as well as the noise.

Wade blowers have been used in numerous performance applications in England, both on the street and for racing. Several companies, including Allard Engineering, have made installation kits in the past to mount them on everything from small four cylinders to the British Ford V-6 and the Rover (ex-Buick) aluminum V-8. However, to my knowledge, no such Wade blower kits are currently being offered, although the blowers themselves seem to be readily available in the U.K. (I have never seen one in the U.S.).

THE PENCO

Do not get this one confused with the Pepco or the Pesco; the Penco is a relatively new Roots-type blower being manufactured in Lincoln, Nebraska. Originally designed in the 1960's by John Camden, a former GM engineer, the Penco is basically a copy of the GM 53-series blower. At this time Penco's are actually fitted with GM 53-series 2-lobe rotors and helical rotor drive gears, although custom-made (though similar) rotors and gears are in the works.

The major differences between a GM blower and a Penco are the Penco's cast aluminum case which incorporates a 4-barrel carburetor mounting flange on the intake side, and 356 T6 aluminum end plates which do not use roller or ball bearings to support the rotor shafts. Instead the end plate is simply reamed and honed to accept the shafts, and fed with a constant supply of pressurized oil from the engine. Thus the end plate itself acts as a bearing surface similar in type (though not in material) to a cam or main bearing in an engine. The end plates must obviously be plumbed with oil from a pressurized engine gallery (an oil line is included in the kit for this purpose); oil is carried from one end plate to the other by a small braided hose, and is then returned directly to the engine valley chamber by a drain passage from the blower through the intake manifold.

The Penco also comes with a long cast aluminum drive snout to allow the small blower to be mounted on large V-8's. The drive is by a single Poly-V serpentine belt and installation kits are designed to spin the blower as much as 200% of crankshaft speed.

Whereas most blowers of this size are intended for smaller engines, the Penco has been marketed so far primarily to owners of motorhomes, large pickups, and other recreation vehicles as a bolt-on cure for the general lack of pep, passing ability, and towing power in these big machines. Consequently Penco installation kits are currently available for big block Chevies, small block Chevies, and big block (383 to 440) Mopars. Installation kits consist of a low profile cast aluminum intake manifold, drive assembly, and oil line hook-up. The kit is designed to use the existing carburetor from the engine (i.e., Quadrajet for Chevies), and Penco strongly recommends, and supplies, a water injection system of their own design to reduce detonation while still allowing the vehicle to be driven on available pump gasoline.

Penco blowers are actually available in two sizes. The "standard"

The standard Penco blower, available in kits to fit a variety of V-8 engines, displaces 140 cubic inches and has a 4-barrel carburetor flange cast into the inlet side. An engine oil line connects to the blower to feed rotor shaft bushings; oil returns to the motor through two holes in the manifold.

model, used on V-8 engines, displaces 140 cubic inches per revolution and produces approximately 4-1/2 pounds boost with the pulleys supplied in current RV-type installation kits. Blower overdrive ratio is 1.75 to 1 on the 350 Chevy, and 2.1 to 1 on 454 Chevies and 440 Mopars. This blower has a flange cast into the top to allow direct low-profile mounting of a Quadrajet, Holley, Carter, or other popular 4-barrel carburetors. The major reason for limiting boost pressure below five pounds, according to Penco owner Don Shoemaker, is to avoid detonation on these big, heavy vehicles. However, the same basic blower (using an earlier case design which does not have an integral carburetor flange) was used by Shoemaker very successfully on a 302-inch small block Chevy engine in a sprint car, making 10 pounds of boost and fitted with a Hilborn injector for running alcohol. This car ran several races with no blower problems. Penco has made numerous custom applications for their blower, including installations as diverse as Oldsmobile and Chevy pickup V-8 diesels, Mercruiser boat engines, and drag racing machines.

The initial Penco blower was a smaller (shorter) 70 cubic inch model which was originally designed with a complete installation kit for 2000cc Pinto four cylinders. This kit is still available, but the smaller blower is also available separately for custom installations (and may be marketed by Blower Drive Service, as well). This smaller blower has an inlet opening about three inches square and requires a carburetor adapter plate.

Recently Speedway Motors, also in Lincoln, Nebraska, has begun marketing the standard Penco blower system for street rods and street machines. We may see more applications in the future.

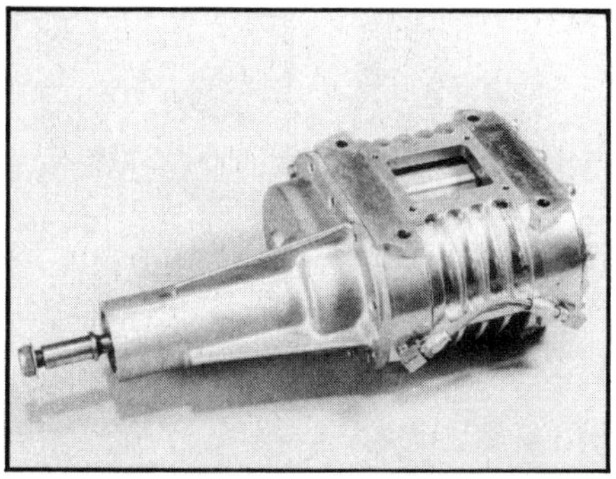

A smaller 70-cubic inch version of the Penco is also available, but not with an installation kit at this time.

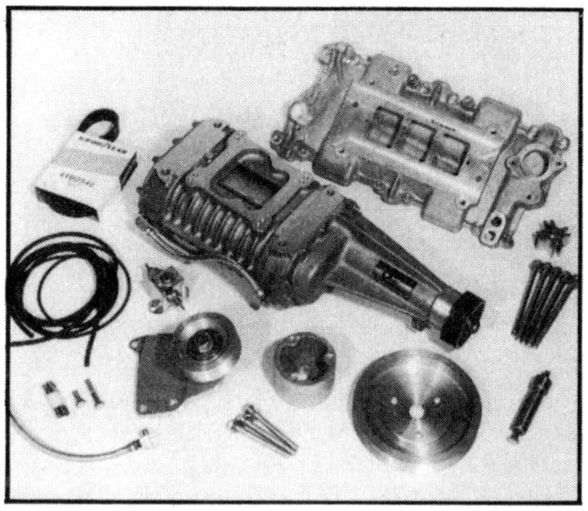

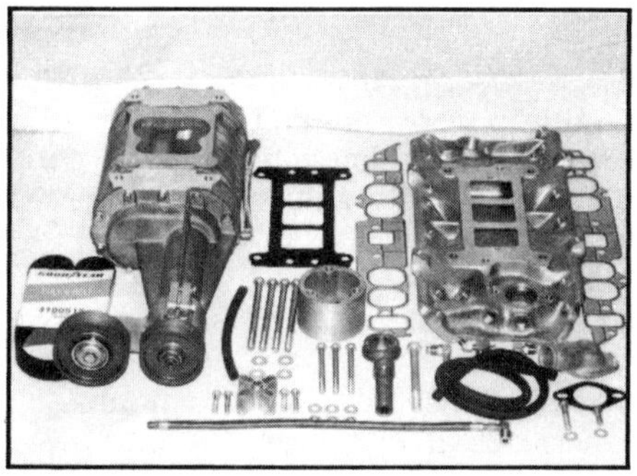

Shown here are complete Penco installation kits for a small block Chevy (top) and a big block Chevy (bottom). Each application has its own low-profile cast aluminum blower manifold, and the carburetor bolts directly to the blower.

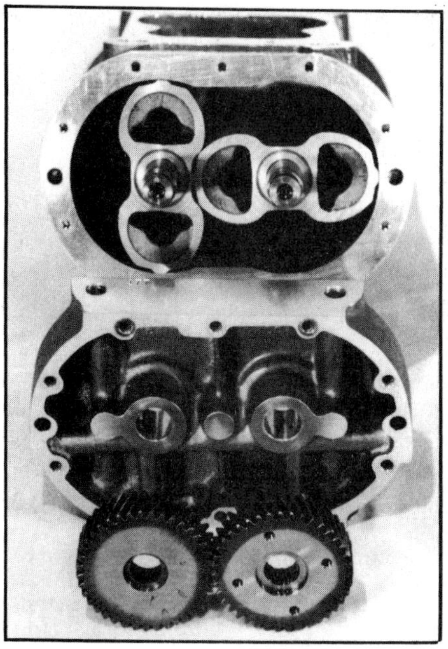

The Penco uses stock GM 4-53 2-lobe rotors and drive gears in a custom housing. A unique feature of this blower is that the end plates use no bearings; instead the aluminum surface is reamed to accept the steel rotor shafts, and is fed a constant supply of oil from the engine.

Penco installation on small block Chevy is simple and compact. The blower is driven by a Poly-V belt, and idler pulley bracket also mounts alternator.

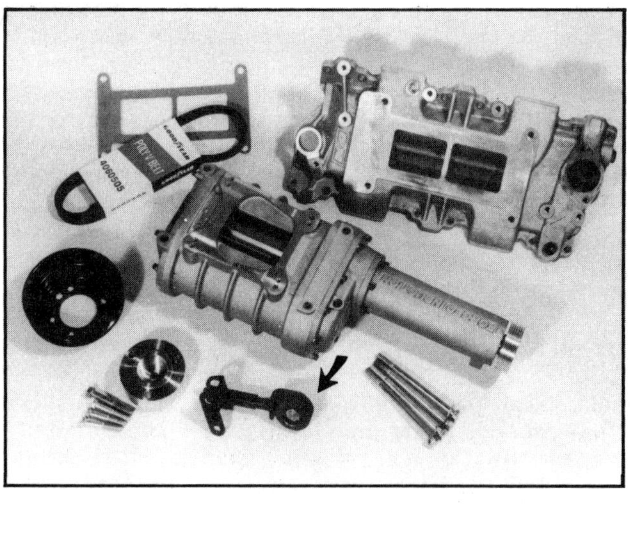

The new B&M supercharger is very similar to the Penco (and the 4-53 GMC), since it was also patterned after the original John Camden design. This is a small block Chevy kit, which uses a spring-loaded idler/tensioner pulley (arrow) to compensate for slight growth of Poly-V belt at higher rpm.

B & M AUTOMOTIVE

Another brand new supercharger just entering the market as this book is being written is the B & M "Forced Induction" blower. Although this unit is a new design and is being manufactured from all new components at B & M, its design concept is very similar to the Penco blower and to the "parent" 53-series GMC. Like the Penco and the Dyers DMS-1 blower kits for Chevy V-8's, the B & M is designed as a low cost and low profile system to give hood clearance in many vehicles. Like these other blowers, the B & M case has a carburetor mounting flange cast into the top which will ac-

Factory test installation of a B&M blower kit on an '83 Camaro required a bubble in the hood to clear the large air cleaner, but everything else bolted in easily. At 7 pounds of boost, this "budget" blower trimmed nearly 3 seconds off quarter mile times on the otherwise stock 305 V-8.

Although the B&M blower is designed to accept a 4-barrel carb, one option is a special intake adapter to mount a pair of Mikuni side-draft carburetors. This photo also shows how the simple drive system mounts on a Chevy.

cept most popular 4-barrels. With the blower installed, the B & M kit raises the carburetor only 3-5/8 inches above the stock location.

The B & M blower uses a cast aluminum case, end plates, and drive snout. Unlike the Penco, it will have ball bearings for the front shaft mount and needle bearings at the rear to support the rotor shafts. The ball bearings are lubed by an oil bath (like a GMC), while the needle bearings are of a permanent lubrication type.

The 2-lobe rotors in the B & M are of the "involute" tip design (rather than a "cycloidal," or round end); this is in keeping with the GMC 53-series design as used in the Penco, the B & M rotors are made from light aluminum extrusions supported on full length steel shafts. The big difference between the B & M and the other 2-lobe blowers, however, is that it will have Teflon insert strips along the rotor's leading edges.

The initial B & M Forced Induction kit will, naturally, be for the small block Chevy V-8. It will include a very low profile aluminum intake manifold and a simple Poly-V belt drive requiring no more than an extra crank pulley and an idler/tensioner pulley. The blower is designed to clear GM HEI distributors. Besides the single 4-barrel mount built into the blower case, B & M will also offer an optional bolt-on mount to accept two side-draft Mikuni carburetors.

Like the Penco, the B & M is designed to be driven at relatively high overdrive ratios (approximately 1.78 to 1 on 350's, 2.0 to 1 on 400's) to produce a maximum of seven pounds

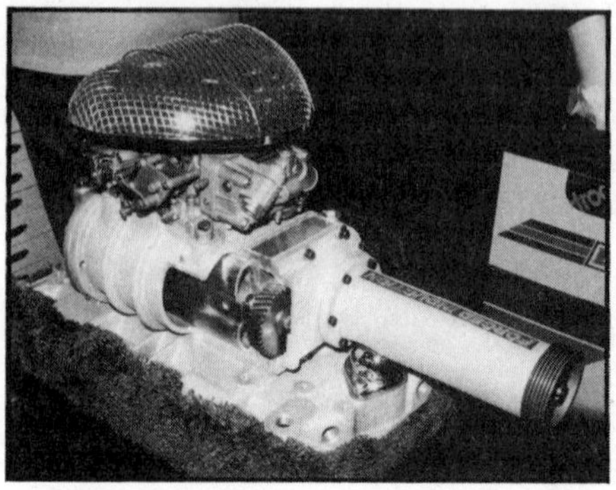

All parts of the B&M are custom made, including extruded aluminum 2-lobe rotors with Teflon-lined tips. Unlike the Penco, the B&M supports the rotors on ball bearings, the front ones being fed by an oil bath.

boost. The reasoning behind this small blower design (besides the cost element and the fact that they wanted a small-diameter blower for a lower profile) is based in large part on calculations derived by Mercedes Benz in the 1930's showing that Roots blower volumetric efficiency increases with blower speed, since leakage around the rotors is a function of time (the faster the blower spins, the less time air has to leak past the rotor clearances). Another factor in the "small blower, high blower rpm" reasoning is that air leakage around the rotors is also a function of the size of the rotors and the case (that is, the total area of the air leakage path). Reducing the size of the rotors and case, while keeping clearances about the same, will reduce the air leakage path.

The fact that these blowers, like the 53-series GMC's, use 2-lobe, straight-cut rotors rather than 3-lobe helical-twist rotors may alter their volumetric efficiency characteristics compared to the more familiar 71-series GMC's (for instance in regard to cavity fill time per revolution). Turning these 2-lobe rotor blowers at high rpm will also reduce the air pulsing effect common to this rotor design, by increasing the rate of pulsing in relation to engine rpm. Some unanswered quantities, however, are the effects of rotor growth due to centrifugal force at high rpm, heating of the air charge from the extra work being done on it (by overdriving the blower), and the increased engine horsepower required to turn the blower at high speed. In the case of the B & M, another unanswered question will be the life of Teflon in the rotor tips after several thousand miles of street use (we really have very little information on how Teflon will live in any type of street blower, including 71-series GMC's, since so few Teflon-sealed blowers have been run any great distance on the street, and then checked for results).

We are sorry that we cannot, at this time, give you more concrete evaluations of some of these blowers. Hopefully by the time you read this the newer blowers (Roots and other types) will have been on the market long enough for some actual road-use feedback to begin either substantiating or disproving the theoretical calculations. In the meantime, it is our intent to give you enough information, or at least raise enough questions, to help you make more informed personal evaluations of these several new superchargers. And while we're on the subject of overall evaluations of currently available street blower kits, we should point out one other efficiency rating that will be a factor in any consumer's ultimate decision: performance per dollar. One of the expressed design features of the B & M blower kit is to manufacture and produce a complete blower system which is very easy to install, for a low price.

KF INDUSTRIES

Another relatively new Roots-type supercharger, introduced in 1980, is the Max 25 by KF Industries of San Diego. Manufactured by Ray Fageol, whose father produced the well-respected Pepco Roots-type blower in the 1950's, the Max 25 is a small-diameter unit designed to give moderate performance increases on stock, small-displacement motors to give them more highway pep, hill-climbing torque, and passing ability.

A unique design characteristic of the Max 25, and a major factor in its relatively low cost, is that both the blower housing and the rotors are made from aluminum extrusions. These extrusions can easily be cut to size before final machining, allowing the blowers to be made in several different lengths. Actually KF currently makes Max 10, Max 15, Max 25, and Max 30 blowers all from the same rotor and housing extrusions, having lengths of 4-1/2, 6-1/2, 8-1/2, and 9-1/2 inches, respectively. The internal displacement per revolution of KF blowers is 7.436 cubic inches per inch of rotor length. Thus the Max 10 displaces 33.46 cubic inches, and the Max 25 displaces 63.2 cubic inches per revolution.

The KF Industries "system" approach is to size the blower to the engine, rather than to overdrive or underdrive one or two sizes of blowers with different pulley ratios. The typical drive ratio on a Max 25 blower is 1.5 to 1 overdrive, and each blower system is designed to produce five pounds of boost at wide open throttle on a given engine.

The Max 10 is designed for engines up to 1300 cc's, and a kit is available to adapt this blower to air-cooled VW's using the stock intake manifold. The Max 15 will handle engines in the 1800-1850 cc range, but no kits are currently made. The Max 25 is for engines up to 2500-2600 cc's, and KF makes a kit (including a cast intake manifold) to mount it on 20R and 22R Toyota four cylinder pickups (it will also fit on Celica passenger cars, which use the 22R engine, but a remote air cleaner is necessary. The Max 30 was being engineered for an installation kit on Chevy 60-degree V-6's, used in S-10 pickups and blazers, as this was being written. Other kits may become available shortly. Currently KF Industries has been arranging programs whereby their blowers will be installed on vehicles through new car dealers, or dealer suppliers. However, individual blowers, complete kits, or kit components are also available directly from KF.

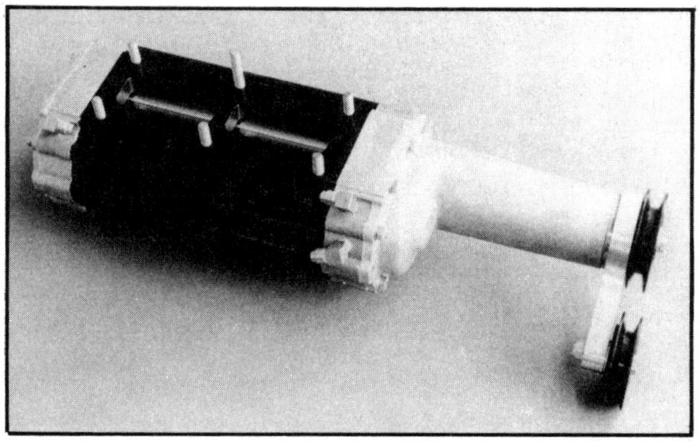

The K-F Industries Max 25 supercharger has been in production for several years, though it has received little publicity. Made by the same family that built the Pepco years ago, it is designed to put out a maximum of 5 pounds of boost, uses a single V-belt drive, and uses engine oil pressure to feed rotor shaft bearings (note oil fittings in end plates).

The KF blower housing measures four inches tall by six inches wide, and is hard anodized after being machined. The rotors are of a 3-lobe, straight cut design, and are fitted with pressed-in and pinned steel shafts which ride on ball bearings front and rear. The end cases of KF blowers are plumbed with a metered flow of pressurized oil from an engine oil gallery, with a small oil line connecting the front and rear blower gear cases. The rotor drive gears are of a straight-cut type, and the blower itself is driven by a single automotive type V-belt.

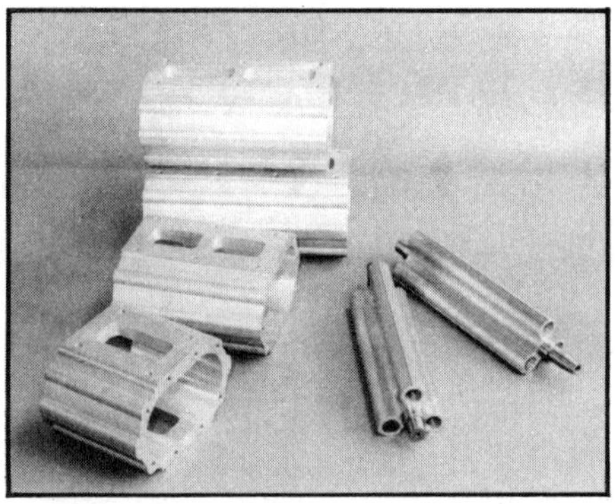

Since both the case and the rotors in the K-F blower are made from extrusions, they can be cut to varying lengths to make different sizes of blowers. After they are cut to size, both the case and the rotors are finish machined.

The Datsun Max 25 blower kit comes with a right angle elbow carburetor adapter to mount a Weber 3236 DGV 2-barrel downdraft carburetor (this particular Weber has double accelerator pumps, which cure a "flat spot" noticeable with other carburetors in this application). The Chevy V-6 installation will use the stock GM staged, computer-controlled, 2-barrel carb atop the blower. For both the Datsun and the S-10, KF has engineered a small add-on computer module, which works in conjunction with the existing on-board computer, to re-program the distributor for a new advance curve compatible with the blower. KF's intention with these complete kits is to maintain a fully operable and certifiable pollution control system in the vehicle, so that the blowers can be installed legally on new cars or trucks.

However, KF also will make what they call a Supermax version of their blower in any of the sizes listed above. This high performance version features specially lightened and balanced rotors, slightly larger clearances to allow higher rpm use, tighter rotor shaft seals, and a unique solu-

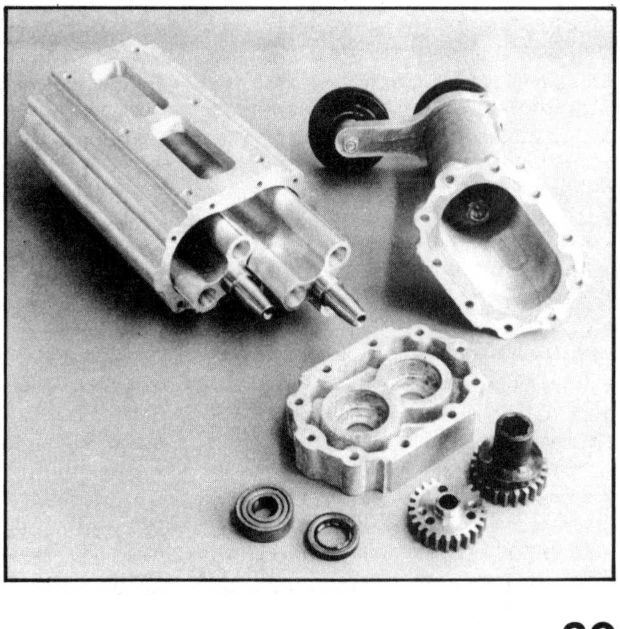

The Max 25 uses cast aluminum bearing plates and drive snout, press-in bearings and seals, straight-cut steel drive gears, and extruded aluminum case and 3-lobe rotors.

tion to the problem of rotor twist at high rpm and boost levels—double drive gears; that is, they add a second set of rotor drive gears at the back of the blower to maintain rotor-to-rotor clearance at both ends. These Supermax blowers should be capable of driven speeds up to 18,000-20,000 rpm, producing boosts into the 20 psi range on small motors. These figures are for short-duration racing, such as drags. The Supermax blowers are built on a custom-order basis, and would obviously require a fabricated toothed-belt drive system.

The primary Max 25 installation kit is for Toyota pickups. As shown here, the blower mounts to the side of the engine on a special intake manifold, and uses an elbow adapter for a 2-throat Weber carb.

Another K-F installation being mocked up in prototype when we visited was this application for a V-6 Chevy S-10/Blazer. It retains the stock computer-controlled carburetor and all smog hook-ups in hopes of meeting 50-state certification.

K-F will also make performance versions of their blower for custom installations, such as the one on this VW-powered dune buggy which uses two side-draft motorcycle carbs, a 1-inch Gilmer belt drive, and produces 14 pounds of boost.

MAGNA CHARGER

Jerry Magnuson of Santa Ana, California, began manufacturing his own, small, Roots-type supercharger in the very early 1970's. Previously he had owned a Chevy six cylinder with a rare 2-71 Jimmy blower, and had drag raced a 4-71 GMC blown '55 Chevy Gasser. But when he initially tooled up to build "Magna Chargers," his intended market was primarily blown fuel drag racing motorcycles. Over the years many of the best drag bikes in the U.S. and abroad have run his blowers.

However, as early as 1973 Jerry had installed one of his little 80 cubic inch blowers on the 327 V-8 in his own Camaro with effective results. A few other automotive prototype installations were made, such as on a VW bug and on a Vega, but the majority of Magnuson's blowers were made for motorcycles until fairly recently. The main reason that automotive kits weren't developed sooner was that, in the first place, the motorcycle business was very successful (the motorcycle market in general was booming in the 70's), and consequently, this did not allow time nor incentive to make models of his blowers big enough to handle medium to large displacement engines effectively.

Yet, during more than a decade of development on his basic blower design, Jerry has been continually refining the Magna Charger, and this is where he has a decided advantage over most of the recent competition. His current blowers are now in what Jerry calls the seventh generation of development.

More importantly, Jerry began with

some inventive ideas about how a Roots blower should be designed. To be quite honest, all other Roots blower builders, past and present, have done virtually nothing to change basic Roots blower design concepts established in the 1930's. GMC blower rebuilders change clearances, add Teflon or Nylon sealing strips, and lately have begun lengthening the rotors and cases, but they all still use basically the same rotor and case design (one or two have played with custom-made rotors, but the exceptionally high cost of casting and/or machining rotors is the major deterrent to change). The designers of other Roots-type blowers for the automotive aftermarket have primarily copied GMC or earlier (i.e., European) designs, in most cases making changes only to lower production costs or to make the overall package smaller.

Magnuson, on the other hand, began with the premise that the low overall efficiency of the GMC 71-series blower was due primarily to excessive turbulence of the air passing through the blower, caused in large part by the twisted rotors. The twist of GM rotors not only "propels" the air toward the front of the case rather than pushing it straight through to the outlet, but it also requires looser clearances to allow for rotor "untwist" at high speed and boost, plus it requires a smaller inlet and outlet port size to maintain rotor-to-case seal. Further, Jerry noted, the helical-cut GM rotor drive gears tend to push one rotor forward, the other back, which alters the rotor-to-rotor clearance setting "up the middle," requiring further increase of this clearance for high-speed, high-boost operation. Not only does this "loosening" of the rotor-to-rotor clearance lower the GMC's volumetric efficiency, but it also allows heated, pressurized air to escape from the manifold up through the middle of the blower, to be propelled by the rotors directly into the incoming airstream.

So the first priority of Magnuson's design was to straighten both the rotor lobes and the drive gear teeth, thus allowing the small blower to run considerably tighter clearances. Another problem with GMC's and other blowers at high rpm is rotor "growth" due to centrifugal force (the lobes tend to stretch outwards), which necessitates wider rotor-to-case clearances for high-rpm blower operation. To alleviate this problem Jerry cast his rotors with support ribs down the middle, and has experimented with several different rotor designs and materials.

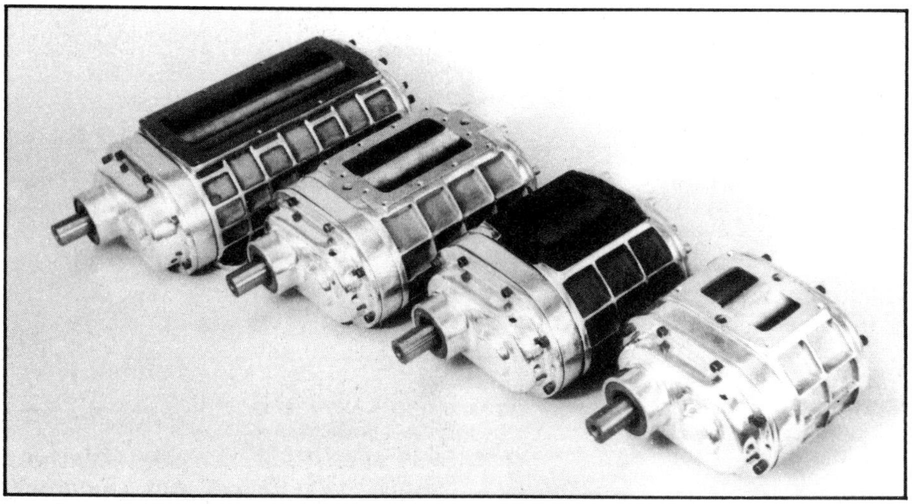

Jerry Magnuson has built his Magna Chargers in several sizes and several styles over the past several years. The little 40 cubic inch size (front) was made for small motorcycles and is no longer in production. The 60 cu. in. and 80 cu. in. models are also basically for motorcycles, or small street engines in the 1200-2400 cc range. The 110 cu. in. version works well on 4-cylinders, V-6's, and small V-8's. Note air divider bars in the inlets of each of these blowers.

But the major design difference in Magnuson's blower is that both the inlet and the outlet ports have been significantly widened compared to other designs. In a Roots blower with two-lobe rotors, the case wall must maintain a seal with the rotor tip for at least a 180° circumference. Otherwise, as the lower tip of the rotor opens at the outlet port, allowing manifold pressure to rush in and pressurize the air trapped in the rotor cavity, this manifold pressure could push the air back out the inlet port, if it weren't sealed by the other rotor tip. Therefore, in the case of 2-lobe straight-cut rotors, the inlet and

Magna Chargers use 3-lobe cast aluminum rotors, and Jerry has improved the design several times over the years. Shown are three "generations" of rotors, with those on the right being the current design.

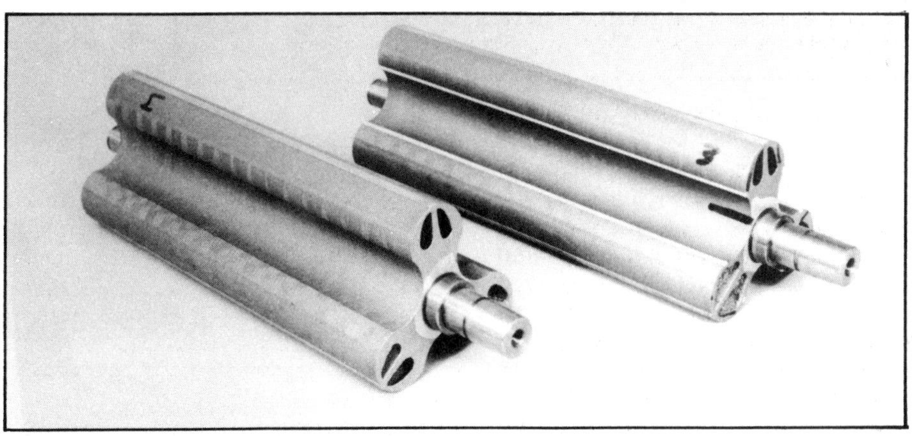

The straight-cut, 3-lobe rotor design is basic to the overall design of the Magna Charger. These blowers can also be set up with or without Teflon strips (the rotor on the right has been fitted with Telfon).

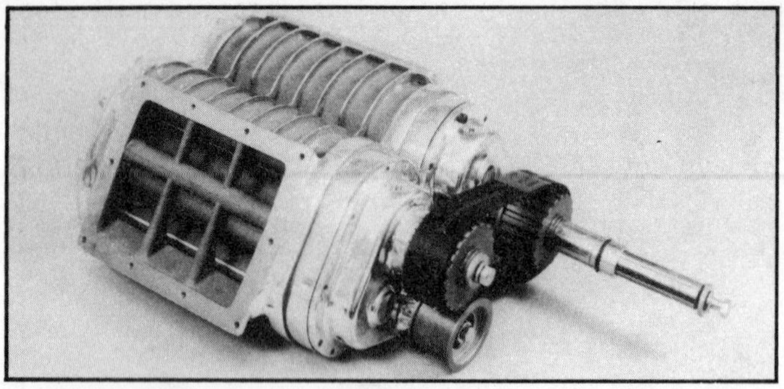

Jerry's latest innovation is his MC-220 supercharger for V-8's, which actually incorporates two 110-inch blowers in one case. The short belt shown drives the two pairs of rotors, while the longer shaft is driven by another pulley and belt which connects to the crank. Note the extremely wide inlet openings in this case (the air divider bar is offset towards the top on this model).

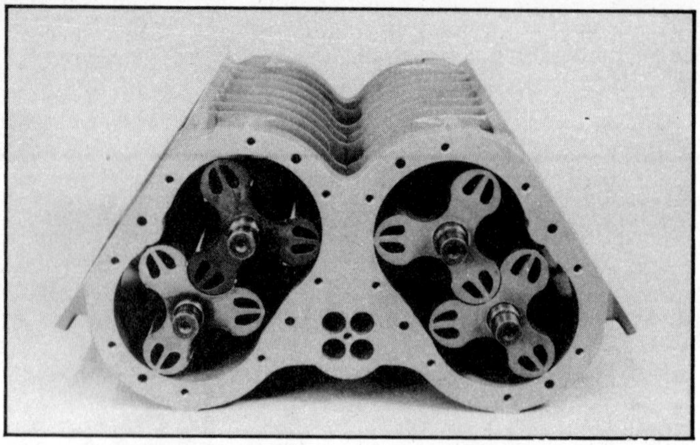

Layout of the two sets of rotors in the MC-220 is apparent from this view. The four small holes in lower center of the case are for the pop-off valve.

outlet ports of the blower can be no wider than the distance between the centerlines of the rotor shafts.

If the blower has 3-lobe straight-cut rotors, however, the sealing circumference of the case wall need only be 120 degrees, thus allowing a considerably larger blower inlet port, outlet port, or both. If the 3-lobe rotors are twisted, however, as in the GMC, then the sealing circumference of the case must be increased by the degree of twist from front to back of the rotor. In a 6-71 GMC, the rotors require approximately 180° case wall sealing, and in some of the new "10-71" through "14-71" elongated race blowers, the rotors do not actually maintain a complete seal with the case wall, and rely on the momentum of the moving air at higher speeds to keep manifold pressure from forcing air backwards through the blower. This effect is similar to camshaft intake and exhaust valve overlap; but, similarly, it yields low volumetric efficiency at low speeds.

According to a patent application paper prepared by Magnuson: "A Roots-type supercharger typically runs at a maximum speed of 15,000 to 18,000 rpm [this is obviously considerably faster than big 71-series GMC's are run on the street, but it is in the range that many of the newer, smaller blowers are being turned], with the air velocity at the inlet typically of the order of Mach 0.33. At such velocities, abrupt cyclic changes in the direction of motion of portions of the moving air may be expected to waste considerable energy in turbulence. Such cyclic fluctuations are inherent in the nature of a Roots blower. While the outer portions of the rotors are pumping air away from the inlet as intended, the inner, meshing lobes are pushing air back toward the inlet and to one side or the other, creating turbulence."

To reduce this multi-directional beating of the air in his Magna Chargers, Magnuson (1) reduced the "timing angle" (the angle of rotation during which the rotors seal with the case walls) from the typical 180° to approximately 125°; (2) in doing so, he enlarged the inlet opening considerably, so that the inlet is larger than the outlet by approximately the same ratio as inlet to outlet air pressure; (3) he thereby, "rotated" the rotor sealing areas away from the top center of the case, and toward the outlet; and then (4) placed a "flow control bar" across the middle of the inlet port of the blower. This "divider bar" is not unique to Magna Chargers, having been used on some 53-series GMC's. Its purpose is to block the incoming air flow from the upward-moving rotor lobes at the center of the inlet port, diverting it sideways to be picked up by the rotor lobes as they begin moving in the same direction as the air flow. Jerry has experimented with various sizes, shapes, and locations of divider bars, and has found a basically round bar to work best.

In other words, what Jerry has done is to widen the blower inlet approximately 100% both to let more air in, and to increase cavity fill time (remember, there's no pressure beyond atmospheric to push air *into* the blower; the rotors will take care of pumping the air *out* of the blower, so there's little advantage to enlarging the outlet). By moving the sides of the inlet port away from the middle, he can not only get the rotors to "grab" the air at a better angle to pull it down through the case, but he can also move the airstream away from the meshing rotor lobes which are moving upwards in the middle of the case, which would otherwise be batting air backwards into the incoming airstream (this is particularly a problem on blowers with straight-cut, rather than twisted, rotors). By placing the divider bar directly above the

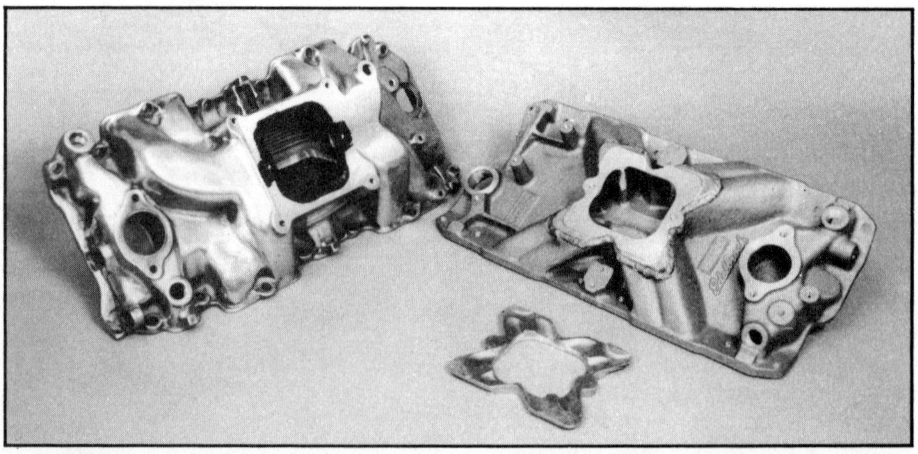

The MC-220 blower for small block Chevies mounts on a Torker 4-barrel manifold to which Jerry welds a special cast flange. He has chosen this manifold, and uses the relatively small inlet opening, to give good low-end and mid-range response on street motors. At left is a big block Chevy manifold which has been machined to accept the MC-220.

meshing rotors, he not only "shields" the incoming air from the upward-moving rotors, but effectively diverts it to the sides, to help fill the rotor cavities.

Actually, according to Magnuson's white paper, careful placement of this flow-control bar will allow it to make use of both the venturi effect and the Coanda effect to "steer" air from below the bar to the right or left, and around the lobe to help fill the cavity, at the same time that it "straightens" the path of incoming airflow. The Coanda effect is the tendency of a flowing mass, such as air or water, to cling to solid surfaces.

Understanding the particulars is not all that important. The fundamentals of Magnuson's design are relatively simple and logical: enlarge and redirect the inlet to increase the blower's volumetric efficiency, and reduce turbulence within the blower to increase its adiabatic efficiency, as well as streamline the airflow path through the blower. Even the fact that the sealing angle in the Magna Charger is reduced lowers air-to-metal friction and turbulence in the rotor cavity. Basically, what Magnuson has done with the Roots-type blower is analogous to what cylinder head designers and porters do to increase airflow through the ports of a normally aspirated engine. According to tests conducted at Garrett Airesearch, Magnuson has seen overall efficiency increases of about 15% for his Roots design, raising overall blower efficiency to about 65%.

Since Magna Chargers were first made for motorcycles, they come in very small sizes. Jerry started with a 40 cubic inch unit, but it is no longer in production (only about 200 were made). The current line includes a 60 cubic inch, which will pump 20-30 psi on 500 cc race motors, or 7-9 psi on 1600 cc street engines; an 80 cubic inch used on 1000 cc race bikes at 30 psi or on 2500 cc street motors at 7-9 psi; and a 110 cubic inch model which will handle race motors up to 2000 cc or street engines up to 300 cubic inches. A kit is available to mount the 110 blower on a Buick V-6 using a modified Weiand 4-bbl intake manifold with a machined, bolt-on blower adapter plate. No other installation kits for these blowers have been developed as yet, but several have been adapted to small inline engines using a variety of pulleys and other components available from Magna Charger. Blower Drive Service has also been marketing the Magna Charger line and manufacturing custom installation kits for various 4-cylinder and V-6 engines.

The latest development from Magna Charger, however, is a new "double blower" which uses a special case to house two sets of 110 rotors for a total displacement of 220 cubic inches per revolution (making it slightly smaller in capacity than a 4-71 GMC...though reputedly more efficient). It is nothing more than two 110 blowers mounted together in a siamesed case, linked together by one blower belt, while a separate belt drives the whole assembly from the crankshaft. This design allows Magnuson to make a blower large enough to handle popular small block V-8 automotive engines without having to retool for a larger-diameter unit, or make an excessively long blower from the existing components.

At the same time, the new MC 220 double blower, besides being a very impressive-looking piece, is designed as a super low-profile unit capable of being installed under the hood of most any late model vehicle, including new Camaros and Corvettes. At our press time, the only installation kit available for the MC 220 is for the small block Chevrolet. It is a very complete installation package including a modified Edelbrock Streetmaster 4-bbl intake manifold, blower drive, accessory mounting brackets, carburetors, linkage, and air cleaners. For the low-profile system, dual Dellorto 40 mm side-draft carburetors are supplied. If hood clearance is no problem, upright carburetor adapters are available for dual four-barrels.

This set-up is designed to be overdriven about 30 per cent (130% drive ratio) on 350-inch Chevrolets to give approximately 8-8.5 pounds of boost. According to Magnuson, the good adiabatic efficiency of this blower will permit boost levels up to 12 psi on 91 octane gasoline without need for water injection or other anti-detonant substances.

Magna Charger offers the MC-220 blower in a complete kit for small block Chevies which includes the blower, manifold, drive, two side-draft Dellorto carbs, air cleaners, fuel lines, and throttle linkage.

Magna Charger also offers a complete kit to mount the 110-inch blower on Buick V-6's. A cast adaptor plate mates the supercharger to a 4-bbl intake manifold.

CHAPTER 6

OTHER TYPES OF SUPERCHARGERS

CENTRIFUGAL BLOWERS

The only similarity between a Roots-type blower and a centrifugal supercharger is the end result they produce, and even in this respect they are significantly different.

In the first place, a Roots is literally a

Although Roots type blowers have been much more visible on street machines lately, centrifugal blowers were more common on street-driven vehicles in the U.S. in the 20's and 30's...and even through the 50's. You will probably be surprised to learn that there were even "Jimmy" centrifugal blowers, as evidenced by the large snail-shell unit (arrow) on this inline diesel.

pump, whereas the centrifugal is more like a fan, or a literal *blower*. The Roots traps and moves a given volume of air through it with each revolution of the rotors (not accounting for leakage); consequently it is known as a positive displacement pump. Its pumping output (again, not counting inefficiency) is directly proportional to the speed at which it is being turned.

A centrifugal blower, however, is an inertial compressor. Its one primary moving part is a fixed-blade fan called an impeller, which actually looks like a disk with fan blades protruding on one side. Whereas the Roots blower is similar to a typical engine oil pump, a centrifugal blower is similar to a typical engine water pump.

The impeller is enclosed in a housing shaped more or less like a snail's shell, with an inlet hole above the center of the impeller, and an outlet at the end of the chamber of the "shell." Air enters the blower axially (along the rotating axis of the impeller), being sucked in by the low pressure created by the "fan," which is blowing air out of the housing. As soon as the air enters the blades of the impeller, however, it is rapidly accelerated radially, towards the outer tips of the fan blades, by centrifugal force. Here is how one

One big advantage of a centrifugal blower is that it can be mounted most anywhere on the engine (within reach of a drive belt). Paxton blowers, which have been the most common centrifugal superchargers over the past 30 years, are designed to be used only on a "blow through" application, which means the blower pumps air through the carburetor. In this case, the carb is fully enclosed in an air pressure box.

source describes the action of a centrifugal blower: "The air-moving element is a form of bucket wheel which hurls air outward from a central inlet against a collector housing called a volute. As the heavy air is hurled outward from the center of the impeller the partial vacuum thus created at the center pulls more air into the inlet. It's a centrifugal pump."

To get a better idea of the principle on which a centrifugal blower works, think of a merry-go-round of the type found in schoolyards or city parks. One or more children push the merry-go-round, while the rest hang on and spin around on it. If you are sitting near the center of the merry-go-round, you won't be spinning very fast. But the farther toward the outside you move, the faster you go, even though the merry-go-round itself is still turning the same speed (that is, the same rpm). Furthermore, as you may remember from your childhood, the spinning motion of the merry-go-round tends to force your body towards the outside edge (that is, radially), and, if you don't hang on tightly enough as it picks up speed, it will hurl you off the edge with considerable velocity. The children's game of crack the whip operates on the same principle.

The tendency of a rotating body to want to fling objects on it, or contained by it, out to the edges is called centrifugal force. It is a demonstration of the laws of inertia, which can be used in the case of the centrifugal blower almost like an "energy multiplier."

For instance, let's says you and a friend are playing on a merry-go-round. You are pushing, and your friend is riding. If your friend is sitting on the outside edge, it will be hard for you to get the merry-go-round going (a body at rest tends to remain at rest), because you will have to accelerate his weight at the speed of the outer perimeter of the merry-go-round. However, if he moves in toward the middle, it will be much easier to get the thing going (because you are pushing rapidly at the outside edge, but his weight is moving slowly in the center—this is the same principle as reduction gearing: using a low gear to get an automobile moving, for instance, and then shifting to higher gears once it's under way). So, since you are an intelligent kid, you have your friend sit in the middle while you get the merry-go-round underway. Then, once you have it spinning rapidly, you tell him to move out toward the edge, where he'll have the thrill of spinning around at a higher speed, but you won't have to push any harder or faster to keep him going (a body in motion tends to remain in motion).

The reason a body tends to be flung off a rotating object is that inertia (the fact that the moving mass wants to remain in motion) is linear—that is, the body wants to keep moving in a straight line. But the rotating object moves it in a curving line. If the body is firmly attached to the rotating object, its mass will tend to keep the object rotating, but part of its inertial energy (also called kinetic energy—or the energy "contained" in a moving body) will be lost.

As long as the blower shaft is sealed against fuel contamination, a centrifugal blower can be installed in either a blow through or a draw through configuration, which makes it even more versatile. This Mach 2 draws air and fuel through the carb, spins it through the blower, and then pumps it into the manifold.

To illustrate this principle, again think of yourself on the merry-go-round. If you are sitting on the outside edge and your friend is pushing at a good speed, the weight of your body will keep the merry-go-round spinning for quite a while after he stops pushing. This is the effect of inertia, or kinetic energy—work continues to be done although no apparent energy is being put into the system. But in this case, it is only part of the energy available. If you are sitting on the outside perimeter of the merry-go-round, you're going to have to hang on with lots of your own energy to stay in place—an amount of energy equal, but opposite, to the amount of kinetic energy which would propel your body in a straight line (tangentially) off the merry-go-round as soon as you let go.

Perhaps we are overdoing the explanation of how a centrifugal blower works—this is just basic junior high school physics—but it does illustrate why a centrifugal blower is much more efficient than a Roots or other type of positive displacement pump, which simply pushes the air with direct force. By using the "energy multiplying" effect of centrifugal force (the inertia of the mass of the air molecules) to help compress the air, consequently the mechanical efficiency of a centrifugal blower is much higher than other types, meaning that it requires less horsepower from the engine to make good boost.

The spinning blades of the impeller in a centrifugal blower rapidly accelerate the air molecules, greatly increasing their velocity, which gives them a considerable amount of kinetic energy. How this velocity is transformed into pressure is a bit more difficult to understand. If you look at a cross-section of an impeller and its housing, you will notice that most (though not all) types decrease in

The several machined discs in the foreground represent the tedious trial-and-error process of reproducing an impeller for a 1920's Miller supercharger at the Briggs Cunningham auto museum. In the background is the Miller blower housing, which has a definite nautilus shape.

Centrifugal blowers were not only used on several production automobiles in America before WW II, but they also dominated Indy-type racing, usually on Miller or Duesenberg engines. This supercharger is actually a Miller copy (a 1927 Cooper), which uses a straight-bladed rotor and a slightly snail-shaped housing. The inlet is in the center.

An interior view of the Miller blower shows the relationship of rotor to housing. This blower uses a straight-blade impeller, but also has a diffuser ring which extends beyond the impeller tips. It can be plainly seen in this photo.

cross-sectional area from the center to the tip of the blades. That is, the cavity enclosed by a pair of blades and the housing is wedge-shaped, tapering towards the outer edge. It would appear, therefore, that air would be compressed in this cavity as it is "squeezed" outward into the smaller area near the impeller tip. But such is not the case, for two reasons.

First, although the cross-section (or the height) of the blade decreases toward the tip, the distance between two adjacent fan blades, since they are radial, increases from the center to the edge, thus increasing the enclosed volume. The combined effect might be a constant-area cross-section, but in fact many centrifugal housing designs (as used in turbochargers) actually have parallel walls. Second, and much more significantly, as the air is

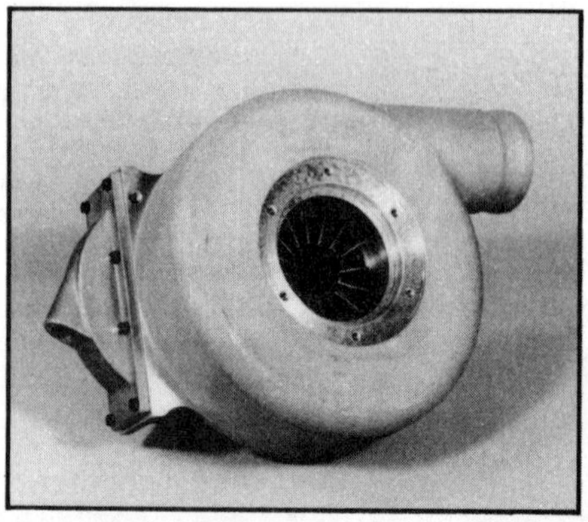

The Mach 2 blower uses a toroidal, or doughnut-shaped housing which has the same cross-sectional area all the way around the circumference. The inlet port is in the center in this view, with the outlet at the upper right.

accelerated along the impeller blade, two things happen to it. It tends to "stack up" against the face of the blade, which would compress the air. But at the same time, as the blade accelerates the air molecules radially, they will "stretch out" as their velocity is increased. This latter condition is similar to the effect of a carburetor venturi which, as you know, decreases the density and the pressure of the air by increasing its volume (the air molecules are stretched farther apart).

All of this may be a little confusing, but don't worry too much about exactly what is going on inside the centrifugal blower housing. The one point we would want to make here is that it is nearly impossible to compute a volumetric efficiency figure for a centrifugal-type blower, since it is not a positive displacement pump. We do not speak of the internal "displacement" of a centrifugal blower, as we do of a Roots or a vane-type blower. In fact, the shape of the impeller blades and the housing have nearly as much effect on the output of centrifugal blower as does its physical size.

The adiabatic efficiency of the centrifugal supercharger is very significant, however. Remember, the energy imparted to the air by the blower (ultimately by the engine) is minimal in a centrifugal unit because of its good mechanical efficiency. That is, there is comparatively little work done on the air—since we are actually using the weight of the air molecules to provide much of the energy—and therefore the centrifugal supercharger does not heat the air nearly as much as a Roots blower does. The centrifugal blower does not "beat" the air, and the flow of air through the blower is much smoother than that in a Roots.

Second, the compression process itself is much more efficient in the centrifugal supercharger. After the air molecules are accelerated along the impeller blades, they are flung off the ends at a velocity approximately equal to that of the tips of the blades. As this high-velocity air is hurled into the scroll housing, it literally piles up against the air already packed into the housing and the manifold, which means that the air actually compresses itself by its own momentum. The kinetic energy imparted to the air molecules by the impeller in the form of velocity is immediately, and efficiently, converted into static energy in the form of pressure. Some heat is certainly generated by this process, since the volume of the air is reduced as it "stacks up" (thus increasing the *density* of the air at the same time that it in-

This view of a Paxton supercharger clearly shows the snail-shape of its housing. The fan blades pull air into the inlet port; the hole at the top is the outlet port.

creases the pressure...which is what we want from a supercharger), but the heat gain in a good centrifugal blower can be so low as to approach 100% adiabatic efficiency. Even the less sophisticated types used for automobile superchargers are capable of 70% to 80% adiabatic efficiency.

Further advantages of the centrifugal supercharger over other types are the extreme mechanical simplicity of the impeller and housing, in which there is only one basic moving part. There is very little stress on the impeller, it is very lightweight, and it does not "mesh" with or come in close contact with any other moving parts. Therefore there is little need for strengthening, clearancing, or ever having to rebuild these components. Also, there is no pulsing of the output airflow from a centrifugal blower, as there is in most other types, nor are there any eccentric moving parts to cause vibration or wear, as in a vane-type, wankel, or piston pump. And finally, a centrifugal blower can produce significantly strong boost pressures from a relatively small and lightweight unit which can be mounted in a variety of convenient locations on the engine.

Is this beginning to sound too good to be true? As you might expect, there are a couple of flies in the ointment—otherwise every car would run a centrifugal blower, and that would be that.

The major problem is that not only is the centrifugal blower not a positive displacement pump, but that the kinetic energy imparted to the air molecules by the spinning impeller increases roughly as the square of the

This Miller centrifugal blower impeller used forward curved tip blades as early as the 1930's. However the tips, which were bent in a jig after the blades were machined, tended to break off under high boost pressures, so the design was not used extensively.

tip speed of the blades. This might sound like another asset at first—if you get ten pounds of boost at a given blower rpm, just doubling the blower speed will give you four times as much boost, and tripling blower rpm will give you nine times more boost.

But think of it in reverse. Your engine can only handle a given amount of boost on the street; therefore the maximum boost level is set. Let's say you want ten pounds of boost at 6000 engine rpm, and you gear the blower accordingly. No problem. But at 3000 engine rpm the blower speed will be cut in half (since the blower is geared in direct proportion to the crank, regardless of the overdrive ratio), and the boost level will be reduced by *four times*, to only 2-1/2 pounds. If the centrifugal blower makes seven pounds boost at 6000 rpm (which is typical for a street set-up), it will only make 1.75 pounds at 3000 rpm.

Thus the centrifugal blower is very "peaky," or speed-sensitive. It will give good, efficient boost at higher rpm's, but very little boost—or torque—at lower engine speeds. As a result, a centrifugal blower is excellent for top-end speed contests, such as Bonneville, or sustained high-rpm racing such as on a long oval track (i.e., Indianapolis). But a centrifugal blower does not work nearly as well as a positive displacement type for road racing, for drag racing, or for any type of driving which covers a wide engine rpm range...and street driving would have to fall into this category.

On the other hand, we are comparing centrifugal blowers to Roots blowers (primarily) here, and this "low end response" argument against centrifugal blowers usually refers to racing situations. On the street, you don't want gobs of horsepower *until* you put your foot into the throttle. The centrifugal blower might not pin you to the seat immediately like a Roots will, but it'll come on like gang busters at the top end. More important it will be far less of a parasite on your engine when you are not in need of extra power, it is much lighter and more compact to carry around under your hood (a major consideration for street driving), and it will be much more efficient when it *is* making boost.

Another drawback to the way the centrifugal blower makes boost is that the impeller must be driven at extremely high rotational speeds to get usable boost pressures. Impeller speeds in the neighborhood of 50,000 rpm are not uncommon. Consequently centrifugal blowers, like turbos, require very high quality shaft bearings and a good oiling system to maintain bearing life at these high shaft speeds. However, the fact that the impeller shaft is not subjected to extreme heat levels, as it is in an exhaust-driven turbocharger, simplifies the situation.

To attain the high impeller speed obviously requires some sort of wide-ratio step-up gearing between the crankshaft and the impeller shaft beyond that feasible with overdrive belt pulleys. Using simple spur gears inside the blower housing to overdrive the impeller shaft has been the typical procedure in the past, and this works fine over a moderate rpm range. However, if the engine is rapidly revved up or down in rpm, the blower impeller and shaft would undergo severe rates of acceleration and deceleration. Early centrifugal blowers, which used relatively large and heavy impellers, would strip the teeth off the drive gears under such conditions. Today's automotive centrifugal blowers use small, lightweight impellers, as well as sophisticated drive mechanisms and shaft oiling systems, to eliminate this problem. Nonetheless, this impeller size limitation—as well as a maximum rpm limit—obviously restricts the ultimate boost levels available from these blowers.

Consequently, the user of a centrifugal blower on a street engine must be careful not to over-rev the unit for two reasons. First, the blower itself has a definite "red line" in terms of rotor and gear rpm speeds. And, since the boost output from a centrifugal blower increases roughly as the square of engine rpm, winding the motor an extra grand could shoot blower boost unexpectedly to a dangerous level, possibly resulting in engine damage. Turbochargers often use a wastegate or other mechanical device to protect

The primary operating component in a Paxton supercharger is this simple, small (it's only about 4 inches across), cast aluminum impeller. The straight-blade design is not very sophisticated, but it is certainly adequate, and reliable, for street applications.

Current Mach 2 blower impellers look much like the early Miller designs, and are completely machined from billet aluminum. In the background is the inside of the Mach 2 torus housing, showing the narrow slit where accelerated air is flung into the "doughnut" where it creates pressure.

A cutaway of a Paxton blower graphically shows the relationship between the impeller cavity (1) and the pressure housing (2). All of the mechanisms to the right of (behind) the impeller are simply drive components, and the large steel balls which step-up rotational speed of the impeller are clearly visible in their race (3) in this photo.

against "overcompression," but most of the current crank-driven centrifugal blowers have no such provision.

CENTRIFUGAL SUPERCHARGERS VS. TURBOCHARGERS

Speaking of turbochargers, these are simply another form of centrifugal supercharger. The compressor section of a turbo is exactly the same; the difference is that the turbo uses another impeller and scroll housing, through which hot exhaust gas is piped, to drive the unit. Thus the turbo takes no crankshaft horsepower from the engine to run. Instead it recovers some of the heat energy normally wasted by the engine in the exhaust to do the work of compressing the intake charge. Combustion heats the exhaust, causing it to expand; thus the exhaust is already pressurized. The drive portion of the turbo works in the reverse of the compressor side—it "absorbs" the pressure of the exhaust gas to spin an impeller. The exhaust enters the impeller radially at the blade tips, and leaves axially from the center. The impeller driven by the exhaust is connected directly, by a common shaft, to the impeller which compresses the intake.

We have already discussed the differences between turbos and crank-driven blowers in general. However, some contrasts between turbos and belt-driven centrifugal blowers—which seem at first to operate in identical fashion—might be enlightening. In the first place, as we mentioned previously, the transfer of heat from the exhaust turbine to the intake impeller along the common shaft is a real problem in the turbo, not so much because of the effect on bearing and oil life, but because of the considerable amount of heat transferred to the intake air charge by the impeller, thus reducing the adiabatic efficiency of the turbo. The belt-driven centrifugal blower does not have this problem. In fact the centrifugal blower has the highest adiabatic efficiency of all types currently in wide use today.

Second, although the belt-driven centrifugal blower is more "peaky" in boost delivery than a Roots-type, it is not as peaky as a turbo. In a crank-driven centrifugal blower, regardless of the percentage of overdrive, the impeller rotation speed is directly proportional to engine speed...even though the amount of boost produced by the blower isn't. In a turbo, however, the "peakiness" is compounded since the rotational speed of the blower impeller depends on the rotational speed produced by the exhaust passing through the turbine portion, rather than on a direct link to the engine. Not only is the output of a turbo not proportional to engine speed (as is the case with any centrifugal blower), but neither is the impeller speed, which doubles the problem. Thus the "sizing" of the turbo to the engine, and the matching of the turbine size and design to impeller size and design, is very critical in a turbo application. Further, a turbo will be greatly affected by camshaft profile, since the cam controls exactly how the exhaust will be released to drive the turbine.

So, while a belt-driven centrifugal blower may not produce as broad a boost curve as a positive displacement blower, it will outperform most turbos in this respect. The centrifugal blower is certainly not as touchy about camshaft profile as a turbo is, and it does not restrict exhaust flow in any way. And, even though a turbo robs no crankshaft horsepower from the engine to operate, a belt-driven centrifugal blower still takes very little engine horsepower, because of its high efficiency, compared to a Roots-type blower.

However there is one area in which contemporary turbochargers are far superior to our current automotive centrifugal blowers—the level of sophistication of impeller and housing design. Turbocharging technology has been advancing continuously and rapidly over the last couple of decades,

Turbochargers are exactly the same as centrifugal superchargers in terms of basic impeller and housing design, and the way that they create boost. The only difference is that a turbo uses a second impeller and scroll unit—attached in "reverse" fashion to the exhaust system—to drive the blower portion.

first in the aircraft field, and then in the automotive and diesel engine industry. One of the primary areas of experimentation has been impeller blade design. Plain radial blades (straight blades which extend outward from the center of the impeller hub, like the spokes of a wagon wheel) are the most basic. Early experiments to improve on this design included radial blades with tips extended upward and curved over near the hub to help draw more air into the blower (as on Miller Indy engine centrifugal blowers of the 1930's), as well as straight blades slightly offset from the hub at the center. More recent experimentation has centered around curved blades, or even curved and angled blades, of various types.

Currently the preferred impeller design for turbochargers is a backwards-curved blade, with the tips actually bending away from the direction the impeller is spinning. This design, though not giving ultimately as high a boost level as other designs, gives a much broader boost curve, at the same time that it reduces the amount of horsepower required to drive the blower as impeller speed increases.

The other area of considerable turbo development has been housing design, especially in conjunction with impeller improvements. The simplest form of housing for a centrifugal blower is a torus, or doughnut-shaped housing, also known as a toroidal housing. A more efficient housing for centrifugal blowers is a volute, or increasing-area spiral, like a snail's shell. Obviously the exact size and shape of such housings could vary considerably, producing differing results.

Our current automotive centrifugal superchargers, however, have undergone virtually no design improvements in these two areas since the 1920's. Both the Paxton and the Mach 2 blowers use simple radial-blade impellers, and the Mach 2 even has a toroidal housing. The major improvement in these centrifugal blowers over past designs is their step-up gearing mechanism, which will survive the rpm fluctuations incurred in today's high-performance automotive engines. But in terms of state-of-the-art centrifugal compressor design, they are rather primitive.

On the other hand, the main reason why these blowers do not use more sophisticated impeller and housing designs is cost. Manufacturing a complicated, curved-blade impeller is exceedingly expensive. To be truthful, for typical automotive street supercharging—which is the primary intent of these blowers—we really don't need anything more elaborate. At least, we probably wouldn't want to pay the price, given the margin of performance improvement we could expect in return, for anything more sophisticated.

Today's Paxton supercharger is basically the same as McCulloch models made in the 1950's, except that all current models use a direct belt drive from the crank, as opposed to a variable-speed drive. This small block Ford application uses a large single V-belt with a spring-loaded idler pulley.

A big asset of the Paxton blower is that it is relatively small and surprisingly light. It can be mounted anywhere on the engine as long as the label faces up (to keep the oil, and pickup tube, at the bottom).

THE PAXTON SUPERCHARGER

The Paxton centrifugal blower is almost as venerable as the GMC. As we mentioned in the section on swapmeet blowers, Robert Paxton McCulloch developed his first centrifugal supercharger for flathead Fords well before WW II. This early McCulloch blower was entirely different from the current model. During the war, McCulloch also developed some Roots-type blowers on government contract for use in landing craft and other special vehicles. One or two of these blowers appeared on race cars, but none was ever produced for aftermarket automotive sale.

The current Paxton/McCulloch centrifugal blower was developed by the Paxton Division of the McCulloch Corporation in 1952, and the first kits were made for flathead Fords in 1953. Although this blower has undergone a few changes in belt-drive mechanisms over the years, the basic design—especially in terms of impeller and scroll housing shape, and the unique planetary ball drive system—has remained virtually the same to the present. The Paxton blower uses a plain radial-blade impeller and a volute-type scroll housing.

Actually only a small portion of the Paxton blower unit is comprised of the impeller and scroll. Most of it houses the impeller drive system, which is accomplished by five large-diameter steel balls which ride against two opposing "races." One race is driven by the blower pulley (which is driven, via a belt, by the crankshaft). A spring-loaded clutch pack, much like that in the engine's pressure plate, applies pressure to the back of the driven race, which is directly connected to the impeller drive shaft. Thus the steel balls are sandwiched between the two races, and the clutch pack applies enough force to maintain a friction drive from one race, through the balls, to the other race. However, the rolling action of the balls between the two races acts like the planetary gears in an automatic transmission to step up impeller speed at a 4.4 to 1 ratio over the blower pulley speed.

This unique design feature of the Paxton blower serves a few important purposes. First of all, it is a considerably quieter drive system than spur gears would be. Second, the ball-bearing drive system absorbs less energy to operate than step-up spur gears would. But most important, this pla-

This cutaway of an early Variable Speed McCulloch blower graphically shows that only a small portion of the unit was the actual supercharger—the impeller and scroll housing (1) located at the extreme left. Just behind the impeller are the drive balls, which are held against their races by a strong clutch pack of springs (2). The rest of the case holds the mechanism which changes the pulley ratio, consisting of an electric solenoid (3) triggered by manifold pressure, and a vacuum/pressure operated diaphragm (4) which pulled the back of the pulley in or out against the tension of the spring in the middle. This area in current Paxton blowers is empty except for the oil reservoir and pump.

These photos show how the VS McCulloch blower actually worked. When the pulley flanges were closed, the belt rode on a larger circumference, and the blower turned slower in relation to the crankshaft (top). When the pulley "opened," however, the belt dropped down, essentially increasing the drive ratio and turning the blower faster.

netary-ball drive design eliminates the major failure of previous centrifugal blowers for high-performance automotive engines—the fact that rapid acceleration or deceleration of the engine would strip the teeth off of conventional drive gears in the blower housing, due to the extremely high rotational velocity of the impeller and its consequent inertia. In other words, a major asset of the Paxton blower is its nearly fool-proof drive mechanism.

Another unique feature of the Paxton/McCulloch blower introduced in 1953 was a variable-speed drive pulley. These blowers were known as the "VS" series (for variable speed). The idea was to compensate for the "peaky" boost curve inherent to the centrifugal blower design by driving the impeller at a higher speed ratio at lower engine rpm, and then "shifting" the blower drive to a lower ratio at higher engine speeds. This feature would not only produce more blower boost at lower engine rpm, or "flatten" the boost curve over the engine operating range, but it would also limit maximum blower rotational speed (to about 35,000 rpm) to safeguard against internal blower damage.

The VS blower shifted gears by means of a special V-belt pulley, which had one fixed flange and one movable flange. The blower was driven from the crank with a single large, wide V-belt. At low engine speeds, the blower pulley would be spread apart, allowing the V-belt to drop down between the flanges to an effective pulley diameter of 3.5 inches, giving a blower overdrive ratio of 1.61 to 1 (in most cases). As engine speed increased, so would manifold boost. An electric pressure switch inserted in the intake manifold would trigger a solenoid in the blower when manifold pressure reached five psi, opening a valve between the manifold and a diaphragm in the blower housing. The diaphragm in the blower pushed against a large coil spring, which in turn pushed against the movable flange of the blower drive pulley, causing the pulley to "close," which would cause the V-belt to ride higher in the pulley groove. Thus the blower pulley would have a larger effective diameter, causing the blower rpm to drop (maximum effective diameter was 5.62 inches, giving a 1:1 drive ratio in most set-ups).

This variable-speed drive system also required a variable-tension idler pulley of some sort to take up the slack in the drive belt as the blower pulley diameter changed. Actually, the idler pulley was mounted on a swing arm held in tension by a large spring. As the system was originally designed, the tension of the idler spring, the tension of the spring between the diaphragm and the movable pulley flange, and the pressure in the manifold were all balanced to each other so that the blower speed would vary with engine speed and boost, rather than just shifting abruptly from one speed to another at a given rpm.

The reason we are telling you all this is because there are lots of these early McCulloch/Paxton VS blowers out there, and (as we mentioned earlier) they can often turn out to be your best swap meet blower buys. After the VS blower kit was introduced for flathead Fords in 1953, similar kits were developed for nearly all the engines of the mid-50's, and one source states that "tens of thousands were sold year after year."

Paxton Products will still make the VS-57 blower (on special order), and they can rebuild any units going back to 1952. According to Paxton's Rick Gattennio, "Unless a guy drops it and breaks it in half, we can fix it." Typical rebuilding of a Paxton/McCulloch would consist of checking the impeller for balance, replacing the ball-drive element, replacing bearings and seals, and a thorough cleaning and repainting. At the time this was written, a complete rebuild cost about $450 (including parts cost for the ball-drive element, of about $300).

If you find an older Paxton/McCulloch blower for sale, the first thing to check is the way the pulley turns by hand. The movement of the blower should be smooth, with no ratcheting, and there should be a discernible drag on the pulley. If you have the equipment to check it, the actual drag should be 25-40 inch-pounds of torque on the shaft at 15 rpm. If the blower is of the variable speed type (with the movable pulley), you should try to get the complete installation kit with it, since fabricating your own drive system to properly balance the idler spring tension to the pulley spring tension—not to mention finding a proper manifold pressure switch—would be difficult. If these parts are not available, however, you can easily convert this blower to operate at a constant drive ratio by removing the pulley, the spring behind the pulley, and the diaphragm, and then installing a standard (non-variable) pulley on the blower shaft.

However, if the older blower still has its original ball-drive mechanism, be warned against driving it at a high overdrive ratio with direct drive pulleys. Remember the early engines these blowers were designed for did not turn much over 4000 rpm. Although the basic design of the Paxton blower has changed little over the years, advances in both machining capabilities and metal technology have improved significantly. The current Paxton blowers feature much closer machining tolerances than the older units—especially in the "roundness" of the drive balls, which is critical—and highly improved metallurgy of components. Thus the current blowers will withstand much greater speeds than older ones—though the new parts can be installed in the older blowers.

If you are fabricating your own installation and drive set-up for a used or new Paxton/McCulloch blower, be *sure* that the blower unit is mounted with the label or name plate directly at the top. Otherwise, the oil pump pickup will not be properly aligned in the oil bath inside the blower housing. The oil bath in all reservoir-type Paxton blowers should be filled with Ford-type automatic transmission fluid (designated "FA"); do not use GM-type auto-trans fluid (Dexron).

A variation on the VS-57 Paxton blower was a model introduced in the late 50's called the DO-59. The "DO" stands for "direct oil," and this blower, instead of having an internal oil reservoir, was plumbed into the engine oil system to be fed a constant supply of pressurized and filtered engine oil. The purpose of this change was to increase the flow and reduce the temperature of lubricant supplied to the impeller shaft and the ball-drive mech-

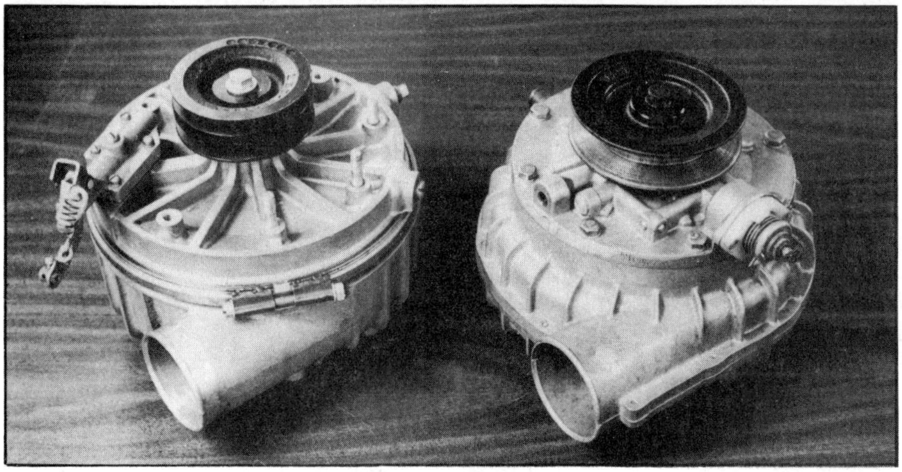

A special Paxton blower, shown on the right in these pictures, was made for factory-option applications on 1957 Fords and Thunderbirds, but only a couple hundred were ever installed. It had an internal system for varying the drive ratio. Even rarer is the larger Phase 1 blower made for NASCAR and Daytona Beach racing '57 Fords, shown on the left.

This is a complete Paxton kit for a smallblock Ford. It is actually a carryover form the old Variable Speed Style (hence the single, wide V-belt and the spring-loaded idler pulley), but it now uses a non-variable pulley on the blower.

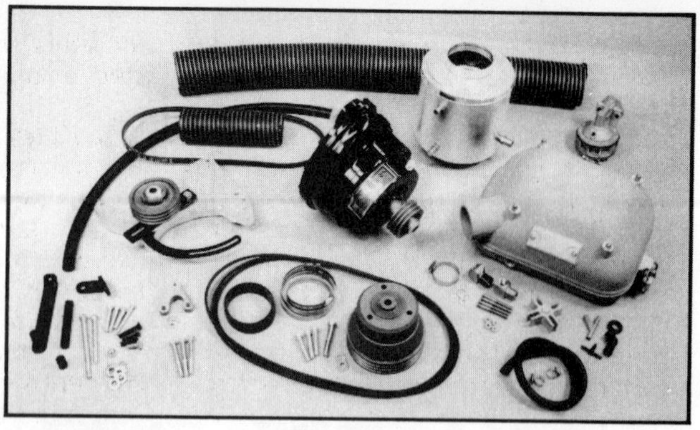

Paxton's mounting and drive kit for small block Chevies is relatively new, and uses three small V-belts to spin the blower. This is a blow-through version, with a complete enclosure for the carburetor. Also note the very large, canister-type air cleaner supplied in the kit, which Paxton says is necessary to get full boost from the blower.

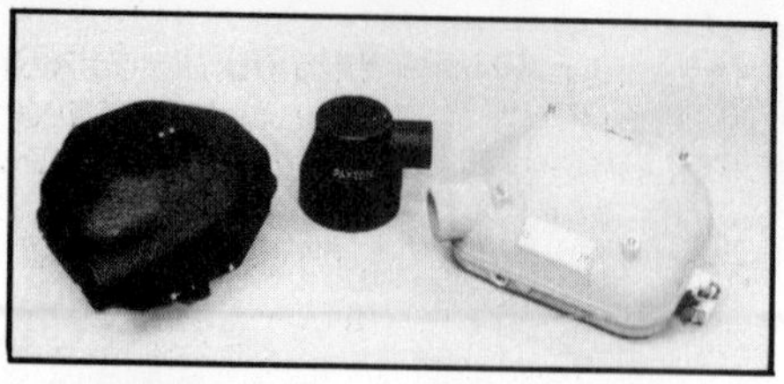

The majority of Paxton installations are of the blow-through design, and they use either a full carburetor pressure box (shown on left and right), or a bonnet (middle) which blows air through the carb throat.

anism. This was intended to prolong the life of these parts in blowers installed on higher-revving engines of the later 50's (such as small block Chevies). However, further problems were encountered with this design because owners either failed to change engine oil often enough, or added substances to the oil (such as STP) which were not compatible with bearings and seals in the blower. If you find one of these "DO" blowers at a swap meet, you can identify it by a large drain outlet in the bottom of the housing, and absence of a dipstick in the top. This design was in production for only about one year because the lubrication problem was subsequently solved by improving the bearings, drive balls, and seals (rather than increasing the oil supply).

Another variation of the Paxton/McCulloch blower was a unit designated the VR-58, which was used exclusively as a factory-installed option on 1957 Fords and Thunderbirds. The "VR" again stands for "variable ratio," but this blower, instead of having a variable-diameter drive pulley, made the drive ratio change internally. Only about 215 of these blowers were installed on production Fords (the option cost about $450 at the time), and replacement parts for them are in very short supply.

A special blower made by McCulloch in 1957 for Ford was a unique unit called the Phase I. It was used very successfully on NASCAR, USAC, Pikes Peak, and other racing versions of 57 Fords. Only 50 of these blowers were produced (if that many, actually) for homologation by the racing organizations. However, after one year of competition both NASCAR and USAC banned the blower, and that was the end of that.

In June of 1958 Andy Granatelli, with his brothers Joe and Vince, purchased the Paxton Products division of McCulloch Corporation from McCulloch. McCulloch had a contract with Studebaker to supply blowers for the Golden Hawk at that time. However, according to Andy Granatelli's autobiography, *They Call Me Mister 500*, Paxton blowers were then suffering considerable life-expectancy problems, resulting not only in loss of general sales, but also of the Studebaker contract after one year. According to Granatelli, he found the problem to be a lack of exact matching of the diameters of the five friction drive balls, as well as a flaw in the design of the blower's oil pump pickup—both of which were corrected at that time.

Granatelli used Paxton blowers to set numerous speed records, primarily in Chrysler products, during the late 50's and early 60's at Bonneville, Daytona, and elsewhere. In 1960 Paxton Products introduced the SN-60 model, which is basically the same unit being produced today. This blower is virtually the same as prior models except that it has no provision for variable drive ratio, either internally or at the pulley. Consequently the housing of this blower is somewhat shorter than earlier models.

In March of 1962 Studebaker Cor-

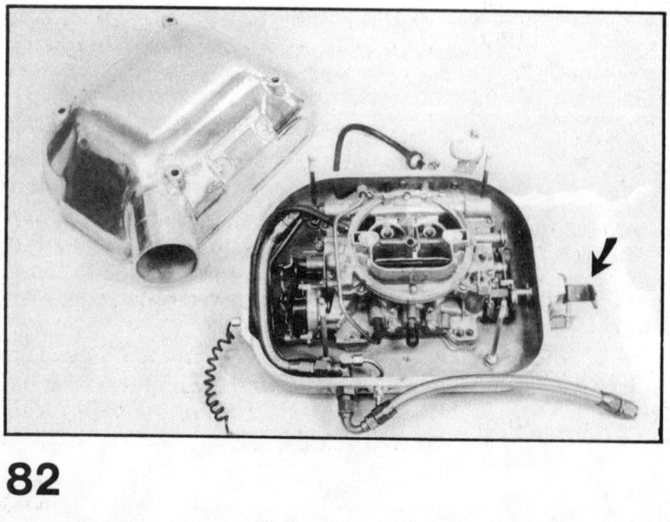

A full carburetor enclosure may not be the simplest, but it is usually the most effective way to set up a blow-through system, since blower pressure is equalized on the inside and the outside of the carb. The air box must be tightly sealed, however, so that no boost is lost, and this means that fuel lines, choke wires or cables, throttle linkage, and other carb hook-ups must be routed through the housing. This particular unit, from Vince Granatelli's personal test car, has extra hook-ups for gauges; but note that the Paxton air box incorporates a pre-sealed throttle linkage (arrow) which attaches to the carb.

Paxton installation on Vince Granatelli's '72 Nova produced 12.21-second/115.68 mph quarter mile times during actual strip testing by Hot Rod magazine. The blower puts out 9 pounds of boost, the car has a 2.73:1 rear axle and Turbo 400 trans, and the 4-bolt 350 is stock except for a Paxton (Engle) blower cam, Edelbrock Streetmaster manifold, headers, an MSD amplifier for the stock HEI ignition, and a re-jetted 625-cfm Carter AFB. Horsepower output with this combination is estimated at 450.

poration purchased Paxton Products, retaining the Granatellis, to produce the SN-60 blower for use on the new Avanti. The R-2 Avanti used this blower with a bonnet over the carburetor, while the R-3 placed the carburetor in a full enclosure. Shortly afterwards, Studebaker placed Andy Granatelli in charge of its Chemical Compounds Division, the primary product of which was STP. Vince and Joe Granatelli remained at Paxton Products.

Paxton is now a privately-owned corporation with Vince and Joe Granatelli as officers. The company continued to produce automotive superchargers until approximately 1972-73, but the major output of the company shifted to industrial blowers after that time (Paxton blowers are used in most Navy submarines for ventilation, and in many industrial applications either as vacuum or pressure pumps).

In 1979 the company reintroduced its small block Ford (289-302) supercharger kit because of popular demand. This kit remains in production using the SN-60 model blower and a single, wide V-belt drive with a spring-loaded idler pulley, similar to the drive used on the VS model blowers. The following year Paxton introduced a new Chevrolet small block installation kit using three small (7mm) Poly-Flex V-belts for the drive. This kit is engineered to clear the hood on late-model Camaros, but will fit most Chevy engine applications.

Another new Paxton kit, using a blower with special seals designated the SN-60 CHV, was designed to provide hood clearance on late-model Corvettes, and places the carburetor in a draw-through configuration. *This is the only Paxton blower designed to work in a draw-through mode.* Do not try to build a draw-through system with any other Paxton blower, since the fuel passing through the blower will seep past the seals and contaminate the oil, or else the flow restriction of the carburetor ahead of the blower can cause the blower to suck the oil out of the reservoir, into the impeller housing, causing the ball drive and bearings to run dry and self-destruct.

Paxton will admit that even the special SN-60 CHV draw-through set-up is not perfect, and that the extra plumbing of the draw-through system lowers the ultimate boost available in the manifold. Paxtons appear to be affected by the length, diameter, and most importantly the number and severity of bends in the ducting between the blower and the manifold. A single 90-degree bend in the ducting between the blower outlet and the carburetor box can result in a one pound loss in air pressure in the manifold.

It is also imperative with a Paxton blower to use the large, low-restriction air cleaner that comes with the kit, rather than installing a smaller aftermarket air cleaner. Not only will any restriction ahead of the blower lessen the blower's efficiency, but in the case of the Paxton, it could cause the blower to draw the oil out of the reservoir. Never run a Paxton (or any other blower, for that matter) without an air cleaner, and place the air cleaner where it will receive cool air rather than heated under-hood air. Finally, never run a Paxton blower without the outlet being hooked up to the pressure box,

New Paxton drive system, which uses three 7mm Poly-Flex steel braided V-belts, is demonstrated on this dyno application. In this case, a bonnet is used to route blower pressure into the top of the carb.

Although it had not yet been released by the time this book was finished, Paxton was experimenting with a new high-boost version of their centrifugal blower. Shown on the right in this comparison with the standard unit, the new blower has an obviously enlarged inlet orifice, and a new style of impeller. Externally the housing remains the same, however.

since this type of blower will be damaged unless operated against pressure.

The fact that the vast majority of Paxton supercharger installations are of the blow-through type necessitates certain modifications and considerations in stock carburetion and fuel supply, and we will cover these in a later section on blower carburetion.

At the time of this writing, Paxton kits were available only for small block Chevy and small block Ford applications. However, these blowers have been mounted on other types of engines using universal or custom mounting bracketry and drive pieces. Paxton will install blowers at their Santa Monica, California, facility, including custom installations on various engines. The Ford and Chevy Paxton street kits are designed to produce approximately 7 to 8 pounds boost (on a 350-inch engine) at 6000 engine rpm. However, higher-performance applications are available on a custom-order basis, with about 10 psi being the maximum boost available on a small block from one blower. For more boost, two Paxtons could be installed on the engine, similar to dual turbo set-ups.

POWER CONCEPTS MACH 2

A brand new centrifugal supercharger which entered the market just as this book was being written is the Mach 2, manufactured by Steve Pick's Metal Concept machine shop and marketed by a group known as Power Concepts, both located in Anaheim, California.

Steve Pick is the person responsible for machining the several intricate components of the Donovan aluminum hemi engines preponderant in Top Fuel drag racing. His expertise as a machinist is well known and respected in the industry. More than twenty years ago, Pick became intrigued with the centrifugal supercharger and began experimenting with designs for one that might replace GMC blowers in racing. In the late 1960's he hand-crafted a centrifugal blower that propelled Frank Rupert's alcohol dragster to 210 miles per hour—a top speed seldom equalled by today's best alcohol dragsters...but that blower died in a spectacular explosion one night at Lions Drag Strip (it was the blower that exploded, not the engine).

The new Mach 2 blower is essentially the same design Pick was experimenting with twenty years ago, with a couple of notable exceptions.

Although it was designed over twenty years ago, the Power Concepts Mach 2 blower is a recent entry in the market. This is the mounting side of the blower unit, showing the gear case. The outlet orifice can be turned to any angle by unbolting and rotating the housing. Small fitting (arrow) receives oil line from engine to lubricate shaft bearings.

Cutaway of Mach 2 blower shows step-up gearing at back of case, and the diaphragm-operated boost control valve unit which attaches at blower inlet. The machined impeller is in the middle.

First, it uses a brand new carbon-based shaft seal recently developed by Pick and for which a patent application has been made. Second, it is fitted with special high-speed shaft bearings developed for a NASA space program. These parts are reportedly good for 75,000 shaft rpm and a 20,000 hour lifespan.

The design of the Mach 2 blower is quite straightforward. It is driven by a large diameter pulley which bolts to the stock engine crank pulley, to drive the blower with a 2-1/2 inch wide toothed belt at a ratio of approximately 1.9:1 to 2.2:1 overdrive. The input shaft of the blower drives a large gear with slightly beveled teeth, which in turn drives a smaller gear on the impeller shaft at a ratio of 4.47 to 1. Thus the combined overdrive ratio of the impeller to the crankshaft is in the neighborhood of 8.5:1 to 10:1.

The impeller is of a basic, straight radial fin design. It is 6.25 inches in diameter (which is quite a bit larger than the Paxton; it also has many more fins). Currently this impeller is completely machined from a solid piece of 7075-T6 aluminum billet, although this part might soon be made from a forging. Attempts to make the impeller, with its thin blades, from other substances or by other processes did not give enough strength to the part to keep it intact under boost loads.

The housing of the Mach 2 blower, cast of 356-T6 aluminum, is of a plain toroidal design (doughnut shaped, rather than scroll-shaped). The entire unit weighs approximately 22 pounds.

Although neither the impeller nor the housing of the Mach 2 is of a sophisticated design, this blower appears capable of making plenty of boost—much more than a street engine needs. Remember that the output of a centrifugal blower increases as the square of impeller speed. In order to keep boost levels within a safe street-driving margin, Pick has added a simple spoiler valve to the intake side of the blower housing. This boost control valve is nothing more than a butterfly, of the type used in a carburetor, operated by an air pressure piston. The piston chamber of the boost control unit is hooked to the manifold by a small hose, so that the piston will "read" the same pressure being made in the manifold by the blower. A large coil spring behind the piston keeps the boost control butterfly open until manifold pressure reaches approximately 5 to 7 pounds. Then boost pressure will overcome the piston spring pressure, causing the butterfly to partially close, which reduces air flow into the blower, thus regulating blower output to the 5-7 pound range.

This blower boost control valve serves several purposes in the Mach 2. First, of course, it limits maximum boost and maintains boost at a predetermined level, in much the same way that a waste gate controls the output of a turbocharger. It also reduces drag on the impeller at higher rpm (which also increases geometrically with impeller speed), which both improves the mechanical efficiency of the blower and saves the thin impeller blades from physical damage (at high boost levels the weight of the air can break or bend impeller blades in a centrifugal blower; when fuel is added to the airstream, as in a draw-through blower configuration, this problem is increased).

The Mach 2 blower can be installed in either a blow-through or a draw-through configuration. This is a draw-through set-up on a small block Chevy.

But the most significant asset of the blower boost control valve is that, like the turbo waste gate, it can be used to compensate for the centrifugal blower's inherent lack of boost a lower engine speeds. In other words, the blower can be geared so that it begins to make good boost at a fairly low engine rpm, and then the control valve will keep boost at that level as engine speed increases, rather than letting boost quadruple as engine speed doubles (as would be the case in an ungoverned centrifugal blower). So, instead of making maximum boost at 6000 rpm, and only one fourth as much pressure at 3000 rpm, the Mach 2 can be set up to make 5-7 pounds of boost by 3000 rpm and maintain this boost through the 6000 rpm range—thus giving a broad, flat boost curve and greatly increasing the usable range of the blower.

This sounds great in theory, but there is at least one hitch. In order to get boost up to the 5-7 pound range at moderate engine speed, the blower impeller must be fairly large (which it is in the Mach 2) and it must be turned at a high overdrive ratio relative to engine rpm. That's fine at moderate engine speeds, but even though boost does not increase from that point with engine speed, do not forget that the rotational speed of the impeller and the impeller drive gears is still increasing at a ratio of nearly 10 to 1. Reducing the amount of air being drawn through the blower (by the action of the spoiler valve) will reduce the stress on these parts significantly, but the weight of the impeller and gears, and their consequent inertia, will impose definite physical limits on these parts as they are accelerated to a potential 60,000 to 80,000 rpm or more.

Although some early advertising for the Mach 2 made claims that this blower was capable of 100,000 impeller rpm, producing boost of up to 100 psi, the fact is that current (as of 1983) Mach 2 blowers have a blower redline of 60,000 rpm. *Do not gear the blower to run higher than this speed.* A variety of belt drive pulleys is available from Power Concepts to adjust the overdrive ratio of the blower. If your engine will never be turned above 6000 rpm, you could use 2.2 to 1 overdrive belt pulleys for an overall drive ratio of 9.83 to 1 (since internal gearing is 4.47:1 in the blower), giving 58,980 blower rpm at 6000 engine rpm. If you are building a engine that will twist up to eight grand or more, you *must* gear the blower accordingly, which might mean that you will lose some low-end boost capability. If you are planning to use this blower in any sort of competition mode, and want to make fairly high boost levels at high rpm ranges, you should first check with Steve Pick on the parameters of this blower's physical capabilities. Otherwise, as they say on medicine bottles, "use only as directed."

The boost control valve of the Mach 2 blower is set roughly in the 5 to 7 pound range at the factory, but it cannot be precisely preset. Once the blower has been installed on an engine, with a good boost gauge hooked up to the manifold, you can adjust the boost control valve by disassembling it and either adding large shims (washers) behind the piston spring to increase boost, or cutting coils from the spring to reduce boost.

However, in a blower test conducted for Hot Rod magazine involving the Mach 2 supercharger, it was discovered that the boost control valve might cause further problems. With the control valve set to limit boost to six pounds on a 350 Chevy engine on the dyno, the engine made significant horsepower and torque increases at lower rpm, but became

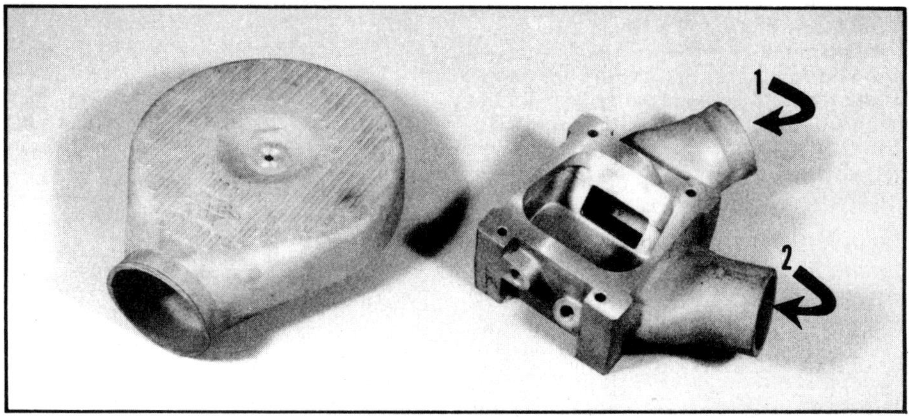

Blow-through set-ups use a simple carburetor bonnet (left), but draw-through systems use the special carb adapter shown on the right, which mounts the carburetor in the stock location on the manifold (to simplify throttle linkage and other hook-ups). However, a floor below the carb seals it off from the intake manifold, and directs air/fuel mix out the elbow (1) which leads to the blower intake (that is, the blower draws the charge through this tube). It then passes through the blower, and is pumped back into the carb mount through the other tube (2), which directs it below the sealed floor, and into the manifold. The Mach 2 carb adapter also incorporates a small flapper door, visible in this photo, which connects the two levels of the housing, so that the engine can draw air/fuel mix directly from the carb into the manifold during low boost/high vacuum engine operation.

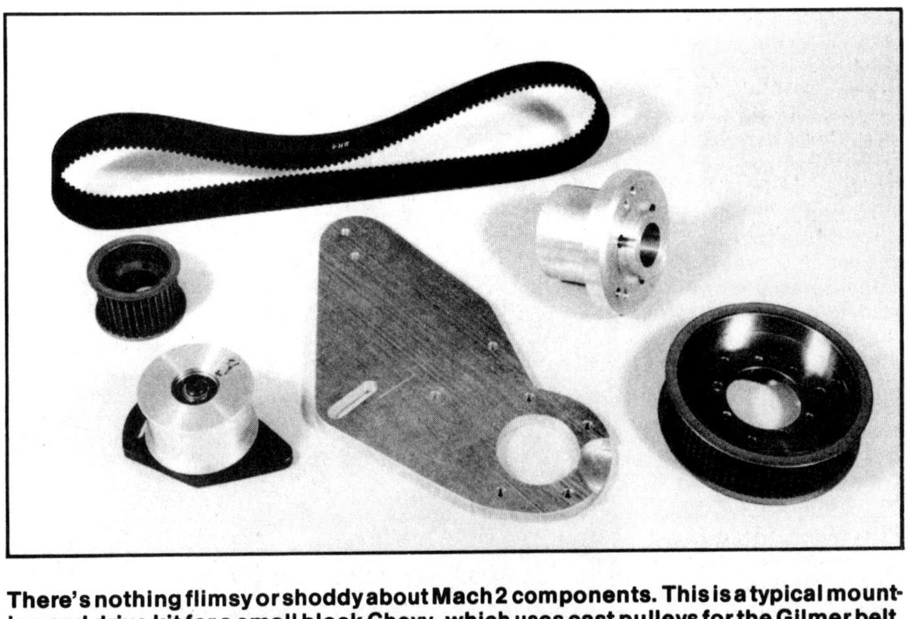

There's nothing flimsy or shoddy about Mach 2 components. This is a typical mounting and drive kit for a small block Chevy, which uses cast pulleys for the Gilmer belt, and a heavy-gauge machined blower bracket. In fact, the big aluminum mounting plate is so beefy, it tends to expand when the engine heats up, and therefore requires that the belt be adjusted loosely when the engine is cold.

This is a Mach 2 installation on an '83 Camaro which uses a hand-made bonnet on a blow-through application to gain hood clearance. Things are admittedly tight here, especially with all the modern engine accessories and controls in place, but the whole system does fit under the hood.

completely unstable above 4000 rpm, failing to hold the load on the dynamometer. This test was conducted shortly before this book was finished, and Mach 2 could not determine a definite reason for the failure. However we would speculate that the problem at least partially derives from the fact that the Mach 2 blower is large enough to make considerably higher boost than six pounds, so when the the boost control valve closes to limit boost to such a low level, the pressure differential across the blower itself becomes exceedingly great. In other words, this big blower is fighting to draw air through its own limiting valve, and is doing considerable extraneous work to do so. We do not yet know what steps Power Concepts will take to remedy the situation; however, the same situation does not seem apparent on street installations (street driving conditions being quite a bit different than dynamometer loading conditions).

Unlike the Paxton, Mach 2 blowers can be used in either a draw-through or a blow-through arrangement. Kits are available in either configuration for a variety of engines. The draw-through system uses a cast aluminum double plenum housing which bolts onto a standard 4-barrel intake manifold and mounts a 4-barrel carburetor 3-1/2 inches higher than normal. The top half of this housing has a large outlet which is connected to the blower inlet by steel tubing. The blower outlet is likewise connected to a large inlet in the lower half of the plenum. Between the two plenum chambers is a solid floor with a flapper door at one end. This flapper allows the carburetor to feed directly into the manifold (without going through the blower) on cold starts. As soon as the blower begins to make boost, however, it blows the door shut and fuel/air mixture passes through the carb into the upper plenum, through piping to the boost control valve, into the blower inlet at the hub of the impeller, is flung at high velocity into the blower housing where the velocity is converted to pressure, exits the blower housing through another tube, enters the lower half of the plenum chamber, and is thence pumped into the intake manifold. Plenum chambers are drilled for specific carburetors; get one to match the type of carb you will be using. The plenum also has two outlet fittings at the back. The upper one will give vacuum under the carb, while the lower one will give pressure in the manifold. Use the lower outlet to attach the pressure line to the boost control valve.

The draw-through system, although it requires a bit more plumbing, is the less complicated installation in terms of carburetor preparation and tuning. The Mach 2 blow-through installations are designed primarily for applications where hood clearance is at a premium, such as in late model Camaros and Corvettes. Rather than using a full carburetor enclosure "air box" like the Paxton, they use a "bonnet" on top of the carb. This makes the blower installation no higher than a stock air cleaner, but it does require some special alterations to the carburetor and fuel supply system, which we will outline in the section on carburetion.

Current Mach 2 blower kits mount

This dyno installation on a small block Chevy gives a better view of the Mach 2 mounting and drive system. With the big blower being overdriven nearly 1000%, and then throttled at the inlet to limit boost to 5-7 psi, the Mach 2 created considerable manifold heat and was probably going into a surge state at higher rpm's during this test. Power Concepts was still working some of the bugs out of this new blower design when this book went to press.

the blower unit to the front of the engine with a large, thick, machined aluminum plate (which varies in size and shape depending on the particular engine and application). It has been found that this plate will expand significantly when the engine warms up. Consequently the blower belt must be installed with sufficient slack to allow for the growth of the mounting plate. When installed on a cold engine, the 2-1/2 inch belt should be loose enough to be able to be twisted 1/4 turn.

The Mach 2 blower is fed a constant supply of oil by a line from the engine. Attach this oil line at a point where engine oil pressure is maximized—usually by attaching a "T" fitting at the engine oil pressure gauge sender orifice. Although the centrifugal blower will not subject the oil to extremely high temperatures like a turbocharger will, engine oil should still be kept extremely clean and changed regularly when the blower is installed on the motor.

Finally, Steve Pick is currently working on a considerably smaller version of the same centrifugal blower which may be in production by the time you read this. The smaller unit will be not only lighter and able to fit into tighter places, but should be capable of considerably higher rpm operating levels. Thus is could be used effectively to supercharge small engines, while still being able to produce street boost for medium and large V-8's as well.

TURBO-RAM SUPERCHARGER

Finally, another new centrifugal automotive supercharger has just been announced by the Drouin Manufacturing Co. of South Windsor, Connecticut. At this time, we have only sketchy information on this unit and have neither seen nor tested a working model. However, it appears that the blower itself is very similar to the Paxton in design, using the same type of impeller as well as the same type of friction-ball drive for internal step-up gearing. The major differences appear to be the large, smooth-angled, cast aluminum carburetor pressure box which attaches directly to the blower outlet port, and a toothed-belt blower drive system.

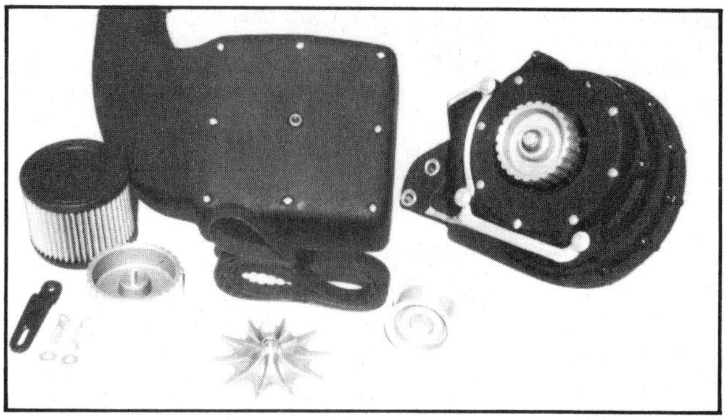

A new centrifugal blower being manufactured on the East Coast is the Drouin. It is obviously similar in design and operation to the Paxton, but uses a different housing and air box, plus a Gilmer belt drive.

Initial announcements state that this new blower will be available for many Chrysler, Ford,and GM production models; however, as expected, the first units will fit small block Chevy V-8's.

THE LATHAM AXIAL FLOW SUPERCHARGER

We discussed the Latham Supercharger, manufactured in Florida in the 1950's and 1960's, in some detail in our chapter on early blowers. Although many racers and automotive experts thought very highly of the Latham Axial Flow compressor, it had one major drawback, and two relatively minor ones, which finally halted its production.

The big problem, as you might guess from our previous description of this blower, was the elaborate machining and assembly process involved in the Latham's design and construction. It was an exceedingly expensive unit to make, leaving little or no profit for the manufacturer. The minor problems were in the high-overdrive belt drive, which was difficult to keep aligned and to keep from slipping, and in the design of the blower intake. Initial Latham blowers, although they mounted as many as four or five carburetors, had very narrow inlet passages into the blower housing, as well as poor-flowing carb manifolding. On later Lathams the intake passages were machined larger, and the front of the fan drum was turned down, to allow better breathing, but the problem was still not fully cured.

Just recently, however, all of the remaining Latham Supercharger parts, patterns, and the right to manufacture the blower under this name were purchased by Richard Paul of North Hollywood, California. Paul, a former drag racer, Bonneville racer, and Formula 1 competitor, will continue to call the revived operation the Latham Supercharger Company. At present he has enough components to assemble approximately fifty superchargers from original parts, and he has installed a prototype (using an original blower) on an '83 Camaro.

However, Paul plans to make investment castings of the rotor and stator fan rings in order to significantly reduce production costs. At the

This Latham supercharger installation on a small block Chevy in a 1934 Ford street rod makes a very impressive sight. However, even though the owner has fabricated his own log-type adapters to mount four Carter side-draft carburetors on each side of the blower, they very likely will not increase the performance of this supercharger since the major restriction is at the small inlets into the blower, which are located on either side at the front.

The Latham blower operates much like a jet engine. A series of curved fan blades on the rotor accelerates the air drawn in through the inlet ports (arrow), while opposite-curved blades on the inside of the housing (stator) act as diffusers to step up pressure. Shown here are rotors and stators from a couple of blowers. These must be assembled one ring at a time, and the size of the blower can be varied by using more or fewer rings.

same time, he will redesign the blade configuration, using an airfoil shape which would have been impossible to machine, but which will be much easier to cast. The rotors and stators will still be assembled as a stacking of rings, which means that the blowers can be made in various lengths, or capacities, simply by adding more rings. Paul plans to make an 11-stage blower for big block Chevies, and 10-stage units for small block Chevies and Fords to begin with.

The inlet volute of the blower will also be redesigned for increased and smoother flow. Rather than individual Carter YF sidedraft carburetors, as used on most original Lathams, Paul plans to cast single, low-profile blower inlet manifolds which will mount one four-barrel carburetor on one side of the blower. The belt drive problem has been solved by using two new, small 7mm V-belts rather than the single flat belt used in the past. Blower overdrive ratio will be 4.9:1 on 305's and 5.3:1 on 350's.

Initial installation kits for the new Latham blower will be designed for 82-up Camaros, 82-down Corvettes, and new Mustang V-8's. However, at the time of this writing none of these kits, nor the new version of the blower, was near production. Projected introduction date on the new components was January of 1984. Check with Latham Supercharger Co. in North Hollywood for further details.

THE LYSHOLM SCREW-TYPE COMPRESSOR

As we said earlier, there are any number of ways to pump, accelerate, or "blow" air. Another type of supercharger design which we have not yet discussed is the Lysholm screw-type compressor. We will touch on it only briefly here, since to our knowledge no example of this type of blower has ever been used on an automotive engine.

These compressors have been used primarily on large stationary engines, and examples include the Elliot-Lysholm and the Broom-Wade. The reason they have not been made in smaller versions for automobiles is that the machining cost to produce the rotors is prohibitively high, and to be honest, the increase in efficiency just isn't worth the added expense.

The Lysholm is designed somewhat like a GMC Roots-type blower in that it has two counter-rotating, spiral-cut, meshing rotors. However, the principle on which the Lysholm operates is nearly the opposite of the Roots. In fact, the compression process in the Lysholm is more like that in the eccentric-vane supercharger.

The rotors in the Lysholm are different, one having convex "lobes" and the other having larger concave "cavities." Unlike a Roots, which carries air around the outside of the case in the rotor cavities, the Lysholm draws air in at one end of the case, pumps it through the case longitudinally *between* the rotors, and then expels it from the other end of the case. The air is not only pumped, or moved, from one end of the case to the other by the intermeshing screw action of the rotors, but it is also "squeezed" between the rotors to compress it.

The beauty of the Lysholm screw-type compressor is that it combines the advantages of the Roots blower with the advantages of the vane compressor, without incurring the disadvantages of either. That is, it is a positive displacement pump, it has no contacting parts or friction drag other than the rotor gears, it therefore needs no internal lubrication, and it is perfectly balanced like a Roots blower, yet it affords internal compression like a vane supercharger, which greatly increases the Lysholm's adiabatic and volumetric efficiency over the Roots', especially at boost levels above 15-20 psi. It's definitely a good design, it

This is an initial prototype installation of a Latham supercharger on a 1983 Camaro by Richard Paul, who has purchased rights to the blower and intends to market new kits. In this application the blower is driven by small Poly-Flex V-belts, and mounts a single Holley 4-barrel low on one side of the blower on a fabricated aluminum plenum.

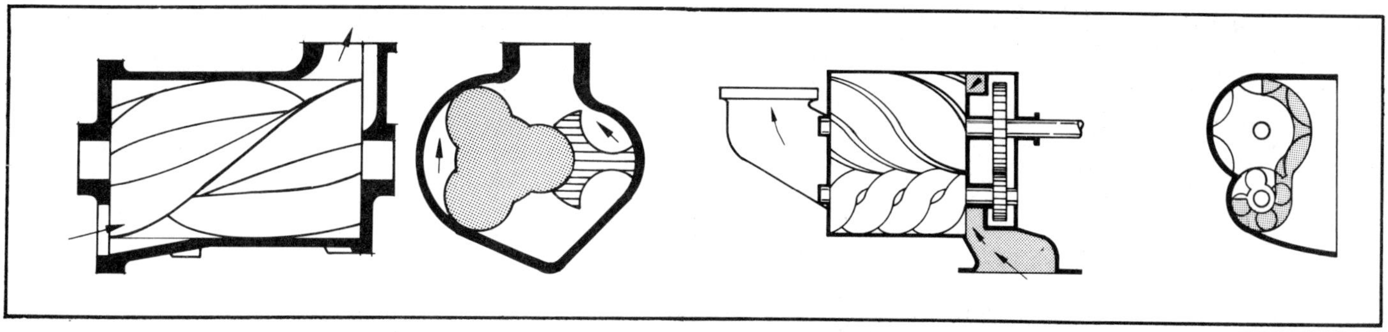

The Lysholm screw-type supercharger is virtually non-existent in this country, primarily because it is so expensive to make. We have no photos of actual examples, but these diagrams portray two variations. The Lysholm is very efficient and effective, being a positive displacement blower with internal compression.

doesn't wear out, it's efficient in all respects—but it just costs too much to machine the two different, spiral rotors to make this supercharger practical, or affordable, for anything but highly specialized applications.

COMPREX PRESSURE-WAVE SUPERCHARGER

A brand new type of supercharger developed and produced by Brown, Boveri & Company, Ltd., of Switzerland sounds extremely complicated, even bizarre. The Comprex pressure-wave supercharger uses exhaust gas energy "pressure waves" to draw in, and then compress, the intake charge in a sort of honey-combed, vaned drum which is rotated by a belt driven from the crankshaft.

In essence, the Comprex blower combines the better characteristics of a turbocharger with those of a belt-driven, positive displacement supercharger. According to Brown Boveri tests, this blower delivers high boost pressure with minimal horsepower drain to drive the compressor (like a turbo), but also delivers immediate response, good low-end torque, and a very broad power curve like a positive displacement blower. However, like most blower designs, it has some drawbacks.

In very simple terms, here is how it works: The rotor, which consists of a series of longitudinal cavities, like a honeycomb, in a rotating cylinder is spun at 1-1/2 to 2 times crankshaft speed by a drive belt. The cavities in the rotor are hollow from one end to the other. At one end of the blower housing a large intake orifice directs fresh air to a large "window," or port. Through this port intake air can enter the cavities in one end of the rotor. Opposite this port, in the same end of the blower housing, is a smaller outlet port and orifice which carries compressed air to the intake ports (or carburetor) of the engine. At the opposite end of the blower is a similar housing and pair of ports which direct hot exhaust gas through the blower. The exhaust enters through a small inlet orifice and port, is fed into the rotor cavities, carried by the rotor approximately 180 degrees, and then emitted through the larger port and orifice to the exhaust system of the vehicle.

You're probably still wondering when and where the compression takes place. Basically, the fresh air enters the rotor cavities, the rotor is turned by the belt to a position past the inlet port, thus sealing off the rotor cavities on that end. However, the exhaust inlet port on the blower's other end opens at the same time, allowing the hot, expanding exhaust gas to rush into the same rotor cavities (that the fresh air is in) from the back end. Thus the pressurized exhaust compresses the intake air in the cavities. The rotor continues to turn, opening the outlet port on the blower's intake side, allowing the now-pressurized air charge to escape, under pressure, into the intake manifold or carburetor of the engine. However, the rotor continues to turn, closing the port before the exhaust gas can escape through that side.

As the rotor continues to turn, the exhaust outlet port is uncovered on the exhaust side of the blower, allowing the still-expanding exhaust to reverse direction and escape out the exhaust exit port into the exhaust pipe. Once this column of exhaust gas has reversed direction and is moving out the exhaust port and down the pipe, the rotor turns to a position which uncovers a scavenge port, and the evacuating exhaust draws in a fresh air charge which is sucked completely through the rotor cavities to both cool the unit and purge any residual exhaust gas from the chambers. Then the scavenge port closes, while the intake port remains open, and the fresh air charge fills the rotor cavities to start the cycle again. The "pressure wave" effect employed to scavenge exhaust and draw fresh air into the blower is very similar to the ram induction effect of tuned intake manifolds and the scavenge effect of tuned exhaust headers.

Thus the Comprex blower uses engine horsepower only to spin the rotor, but the work (or energy) necessary to compress the intake charge is supplied by the hot exhaust gas, just as it is in a turbocharger. However, in the Comprex, the timing and the delivery rate of pressurized air to the engine is determined by blower rotor speed, which is directly proportional to crankshaft speed, as in a positive displacement blower.

However, as we said, there are drawbacks. First, as you might imagine, since the Comprex actually places the hot exhaust and the cold intake air in direct proximity to each other, a considerable amount of heat is transferred to the intake charge (not only by exposure to the exhaust, but also from the rotor, which is designed very much like a heat sink). Brown Boveri claims that heating of the intake charge is minimal since the contact time with intake air is so short; in actuality the important fact is the ratio of compression afforded by the blower compared to the amount of expansion caused in the intake charge by heating it. Secondly, it might be assumed that direct contact of exhaust with intake air in the rotor chambers would lead to contamination of the air charge with exhaust particles. This is supposedly combatted by the purge cycle in the blower. Nonetheless, it is very doubtful that this supercharger could be used in anything but a blow-through configuration (since you wouldn't want to be mixing fuel-air charge with hot exhaust...or you might get the same result as when a backfire "lights off" a GMC).

The Brown Boveri Comprex supercharger has been designed and used primarily for diesel engines—both on large equipment and on automobiles—as an alternative to turbo-

At first the Comprex may seem *complex*, but this cutaway shows that it is actually quite simple in operation. The only moving part is the rotor, which is the drum in the center with the honeycomb interior. The fresh air inlet (1) and compressed air outlet (2) are on one side, while the exhaust inlet (3) and outlet (4) are on the other, much like the layout on a turbocharger.

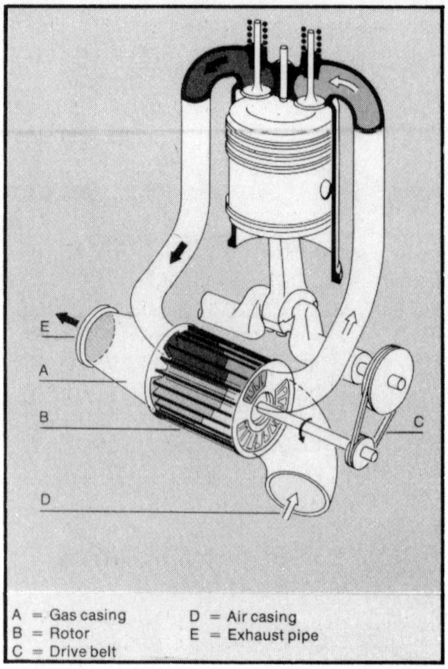

This diagram of a Comprex supercharger shows its layout on a 4-cycle diesel engine (on a gasoline engine the blower would pressurize the carburetor, in a blow-through arrangement). The belt drive from the engine simply spins the rotor to time inlet and outlet pulses for both the fresh air and the exhaust; energy from the crankshaft is not used to compress the air. Instead, heat and sound energy from the exhaust does the actual compressing of the intake charge. Thus the Comprex has an extremely high mechanical efficiency rating. However, its adiabatic efficiency is hindered by direct contact of hot exhaust with cold intake air.

charging. In a diesel application heating of the air charge, or minor contamination, is not nearly as critical as in a four-stroke gasoline engine. Another drawback of the Comprex—although it actually is a relatively simple, uncomplicated design with only one basic moving part—is that it must be made of extremely high quality metals to withstand exhaust temperatures, thus increasing its cost. In this respect it again compares with the turbocharger.

Whether or not the Comprex pressure-wave supercharger would be practical on a street or racing automotive four-stroke engine remains to be seen. In '81 and '82, Ferrari used the Comprex blower on their Formula 1 cars with some promise, but without conclusive results.

VANE-TYPE SUPERCHARGERS

Although we will discuss several vane-type superchargers in this chapter, I want to make very clear at the outset that none of the blowers mentioned in this section is currently available. The Shorrock from England and the Judson from the United States are both out of production, and are relatively scarce in the used parts marketplace. The Bendix was announced in 1981 as a prototype, was installed on a few factory test cars and experimental models, and received considerable press coverage; however, it never went into production in the U.S. (at least not as of this writing), although it reportedly was being produced in Japan. And the Holley supercharger, which is actually a Wankel design, was only in preliminary planning and testing stages; it is not likely to be produced unless an OEM contract is secured with one of the major auto manufacturers for installation on new cars.

The reasons we are covering these superchargers are (1) that the vane-type blower layout should be of interest to anyone studying superchargers; (2) several of these types of blowers have been built through the years, and some are available used; and (3) at least two of the brands discussed in this chapter may become available to the public at a later date.

First let's look at the design of a vane-type blower. The simplest type consists of a cylindrical housing, inside of which a smaller cylinder rotates on an off-center axis. Through slits in the surface of the smaller cylinder protrude any number (usually three or four) of thin, flat blades, or vanes. These vanes are linked to, or ride on, a shaft located at the center of the *outer* cylinder, and are able to slip in and out of the inner cylinder, or "rotor." Thus, as the inner cylinder rotates eccentrically within the housing, the vanes rotate concentrically, maintaining either a slight clearance from, or very light contact with, the wall of the outer cylinder.

The outer cylinder, or blower housing, has an intake port and an outlet port, which are no more than openings in the case wall. As the inner (eccentric) cylinder and vane assembly rotates within the case two things happen. The vanes draw air in through the intake port, moving it much like a paddle wheel moves water. However, as a given quantity of air, or air/fuel mix, is drawn in and trapped between the two vanes, and as it is carried around the blower wall, the inner cylinder begins to move closer to the outer cylinder, thereby "squishing" the air between the vanes. Thus the air is compressed within the blower before it reaches the outlet port, where it is emitted to the manifold.

The vane-type blower is literally both a pump and a compressor. It is a positive displacement pump, like a

Roots blower, since it delivers (moves) the same quantity of air per revolution (not counting leakage) regardless of blower speed. However, the Roots blower is an external compression supercharger, meaning that the air is compressed in the manifold rather than in the blower. The centrifugal blower exhibits a degree of internal compression, but it also primarily produces boost in the manifold or intake tubing. The eccentric-vane blower, however, is strictly an internal compression supercharger—the only one of this type used in a practical manner for automotive supercharging—and therein lies its major attribute.

Logic should tell you that this is a more efficient way to compress air and pump it into an engine—considerably more efficient than the way a Roots blower crams air into a manifold. Most vane-type superchargers will have a higher mechanical efficiency than Roots-types. But it is in their adiabatic efficiency that they excel. First of all, there is very little extraneous motion, or work, done in compressing the air. Second, remembering our gas law from Chapter Four:

$$\frac{PV}{T} = \text{CONSTANT}$$

we can see that increasing the pressure while keeping the volume the same will increase the temperature proportionately; increasing both the pressure and the volume will increase

In a very simple type of vane blower, the rotor—basically a drum—rotates on an off-set axis within the cylindrical case. Vanes, which are just flat "paddles" that ride loosely in slots in the rotor drum, are flung outwards against the housing wall by centrifugal force, carrying air between them and compressing it as the drum moves closer to the wall during its cycle. The Judson blower was of this type.

A more sophisticated type of vane supercharger has the same inner and outer cylinders, with the rotor revolving eccentrically. However, in this case the vanes ride about a fixed pole in the middle of the blower, so that they remain a given distance from the case wall as they slip in and out of the rotor. This allows the rotors to maintain a slight clearance from the housing so that they will not wear, and do not require lubrication.

the temperature even more. However, decreasing the volume while increasing the pressure will reduce the amount of temperature increase. If we could reduce the volume the same amount that we increased the pressure, the equation tells us we could actually keep the temperature constant. However, in real life we are lucky to approach simple adiabatic heat rise in a compressor. The eccentric-vane supercharger comes close.

As we said, the high adiabatic efficiency of the vane-type compressor is its strong point. But it also has some drawbacks, as you might expect. In the first place, such a blower is a complex machine to design, to construct, and to maintain. It has a lot of moving parts, all of which must be set at close tolerances. Further, unlike a Roots or centrifugal blower, some of the components in the vane-type (primarily the vanes inside the inner drum; sometimes the vane tips against the outer wall) maintain physical contact and are therefore subject to wear. In some vane type blowers an oil mist is injected into the housing to lubricate these parts, but oil introduced into the intake air stream is certainly not optimal in an engine and will promote detonation in the cylinders.

A second problem with the eccentric vane supercharger not found in other blower designs is that the off-center rotation of the inner cylinder causes an imbalance in the unit. This characteristic limits the maximum rotational speeds of a vane blower.

A third drawback of vane-type blowers is that they do not approach the same pumping capacity per revolution of Roots or even centrifugal blowers. Combined with the fact that they cannot be turned at high rpm because of the out-of-balance rotation of the eccentric inner cylinder, this

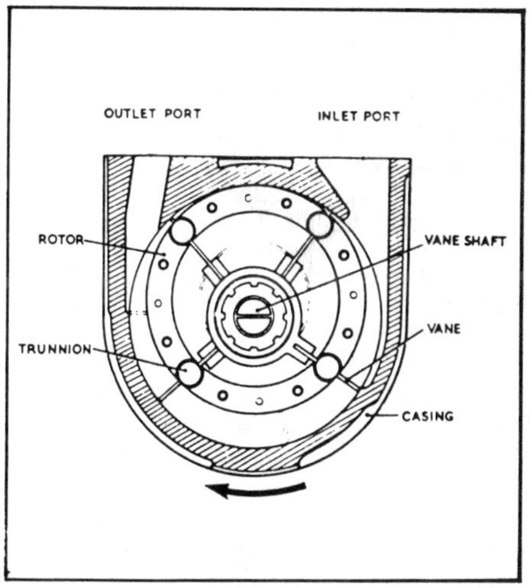

A diagram of the Shorrock blower, one of the more successful true vane blowers, shows the complicated mechanism necessary to hold the vane blades in proper alignment—without hitting the case—as the eccentric rotor spins. The principle for compressing air is very simple, but the machinery isn't.

means that vane-type blowers must be quite large to pump significant quantities of air; yet their physical size is limited by the weight of the off-center rotor. On the other hand, the internal-compression design of the vane blower allows it to make considerably higher boost pressures than external compressors without losing efficiency. Thus it can be used most successfully to supercharge small displacement engines at high boost levels. None of the vane blowers made so far would be adequate to pump a lot of pressure into a large V-8.

THE SHORROCK

The Shorrock vane-type supercharger was designed and produced by Christopher Shorrock in England in

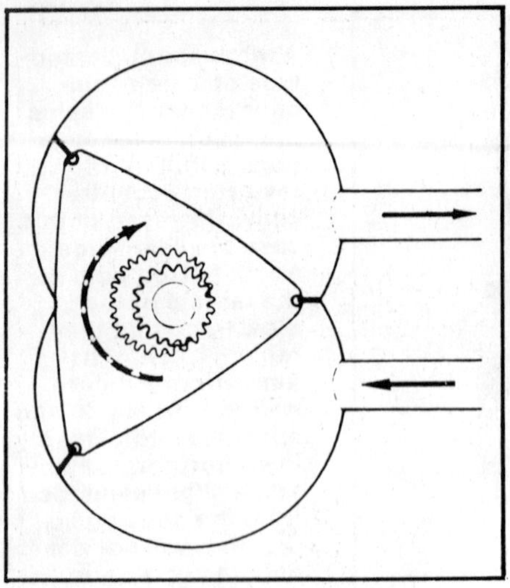

The Wankel rotary engine, as used in Mazdas, is actually a form of vane-type compressor—or, at least it works on the same principle, except that the vanes are fixed at the tips of the triangular rotor.

the 1950's. By 1957 he entered into an agreement with Sidney Allard of the Allard Motor Company to market these blowers and develop installation kits for several European cars and motorcycles. Sidney's son, Allan Allard, continued to market this supercharger into the 1970's, but the units are no longer in production at this point. Apparently these superchargers were relatively common in England, and they were used in several successful drag racing, hill climbing, and rallying installations by the Allards. However, I have never seen one in the United States.

The Shorrock blower uses four vanes which are linked together at the center of the blower by circular rings. These carrier rings ride on a shaft at the center of the cylindrical blower case, supported on ball bearings. The vanes slide through slots in the eccentric inner rotor cylinder, held in place by trunnion shafts. Inlet and outlet ports in the blower case are located approximately 90 degrees apart.

Unfortunately we do not have much technical information on Shorrock blowers. Other than some special "high boost" competition versions, apparently two sizes of the Shorrock were made for street and moderate competition use. The smaller C75B had an internal displacement of 750 cc's per revolution and would handle engines in the 850-1300 cc range at street boost, while the larger C142B displaced 1420 cc's and would handle engines of 1300 to 2000 cc's.

The major drawback to the Shorrock was the oscillating out-of-balance force created by the eccentric inner rotor. This force, or shake, which increases as the square of the rotational speed of the rotor, limits the practical maximum rpm of the C142B to 6500, and of the C75B to 7500, although some of these blowers were wound much tighter than this for short sprints.

Another drawback of the Shorrock was the fact that it required oiling of the internal components. Oil was fed to the blower from the engine oiling system, and metered into the blower case by a small spring-loaded needle-and-seat valve. This oil was in turn fed to the blower bearings and to the vane trunnion shafts, and then allowed to enter the air/fuel stream. Thus the blower used up a small but constant amount of the engine oil supply, and oil was introduced into the intake charge, which increased the engine's tendency to detonate. Also, the small metering valve in the blower required regular checking and cleaning so that it didn't clog and cut off lubrication to the supercharger components, which would result in destruction of the unit.

THE JUDSON

Haddon Judson of Conshohocken, Pennsylvania, began researching superchargers in 1948 and came up with the design of his vane-type blower four years later. The initial unit, known as a model B-210, was made for the flathead Ford V-8. It mounted directly on the intake manifold and had a built-in intake flange for two Stromberg 2-barrel carburetors. However, subsequent models of the Judson supercharger were made to fit various types of foreign or American small cars such as MG's, VW's, Renaults, Mercedes, Triumphs, Sprites, Volvos, and Corvairs. The Judson blowers were sold only as complete kits for these specific cars, and several sizes and shapes of the supercharger were produced for various applications. By the mid-1960's Judson was advertising that, "There are more Judson superchargers in operation than all other makes combined." Supposedly well over 50,000 units were sold before Judson ceased supercharger production in the late 1960's in order to concentrate on the manufacture and sale of transistor ignition systems.

Like the Shorrock, the Judson was best suited to small displacement engines, mostly of 180 cubic inches or less. Boost output for most applications was advertised to be in the 5-7 psi range with the blower driven at, or close to, crankshaft speed by a single V-belt.

The design of the Judson was basically the same as that of the Shorrock. It had a cylindrical outer case with an inlet port on one side and an outlet port of the other, spaced approximately 120 degrees apart. Inside this housing a smaller cylindrical "rotor" revolved on an eccentric axis.

The Shorrock blower was quite compact, the smaller C75B unit (right) being not much larger than an alternator. The larger C142B, shown on the left, has a positive displacement of 1420 cc's per revolution, or about 87 cubic inches.

Four vanes protruded from the eccentric rotor to maintain light contact with the case wall. The only major difference between the Judson and the Shorrock, to this point, was that Judson blowers generally had longer cases and rotors, sometimes of smaller diameter.

The unique feature of the Judson, however, was the design, placement, and composition of the vanes. These blades were inserted in slots in the rotor on an angle, so that the leading edge would be leaning in the direction of rotation. Rather than being anchored to a central shaft, the vanes were held against the case wall by centrifugal force, yet counterbalanced so that they would not rub the case with too much pressure, or friction. Judson called this patented design "balanced pressure" and described it thus: "Due to the weight of the blades and the angle of mounting, the centrifugal force is balanced so that the pressure against the case is kept to a minimum and yet a zero clearance is assured." The big advantage of such a design is that virtually no clearances existed in the blower through which air could leak—thus giving this type of supercharger an extremely high volumetric efficiency.

However, as you might have guessed, there are also drawbacks. First of all, in order to reduce wear on the blower case wall, the vanes were made of "a synthetic material." Although sales brochures stated that "any wear that might take place on the blades is automatically compensated for in the design," and also claimed that blowers had been driven 85,000 miles without need for repair, the fact was that the blades would need replacement after a certain amount of use. This is a critical point to consider if you locate and contemplate buying a used Judson supercharger. Unfortunately, I do not know of a good way to visually inspect or bench test a used Judson, other than installing it on an engine and trying it out. However, I doubt very seriously that you can find replacement vanes for one anywhere today.

The other drawback to the Judson design is that it also, like the Shorrock, requires a constant supply of oil for internal lubrication, and this oil is subsequently introduced into the fuel/air stream. Rather than taking oil from the engine, each Judson blower came with its own oil reservoir (usually a glass jar, similar to those used for windshield washers on early cars). The blower consumed this lubricating oil at the rate of approximately one quart every 1000 driving miles. The oil was sprayed into the blower case not only to lubricate the shaft and bearings, but more importantly to spread a thin film across the case wall on which the vanes rode.

This typical street installation of a Shorrock on a British 4-cylinder uses a pair of V-belts to drive the blower, and a single Solex carb to feed it.

Consequently, as you might expect, the life expectancy of the vanes in a Judson blower depended critically on keeping the oil reservoir filled. With proper lubrication, vanes generally lasted about 20,000 miles, or more. However, any time that the blower was allowed to operate without oil would greatly shorten the life of the vanes, if not destroy the blower.

Despite the fact that so many Judson superchargers were produced during the 50's and 60's, very few show up either at swap meets or on cars these days. Many of the units were shipped to England and Europe, since they were made primarily for foreign cars. But it might not be inaccurate to assume that plenty of them died in action because owners, or subsequent owners, did not fastidiously feed them oil.

As we have said, a supercharger can make a little engine run like a bigger one. The Shorrock added to this 1500 cc British Ford engine by Alan Allard provides plenty of fun in a dragster chassis. Note single-throat fuel injector and Gilmer drive.

Although reputedly thousands of Judson vane-type blowers were sold throughout the world in the 50's and 60's, hardly any remain today, probably because people ran the oil mist reservoir dry and wore out the vanes. This swap meet Judson, made for a Corvair, does not have the oil jar and lubricating system with it, you will notice.

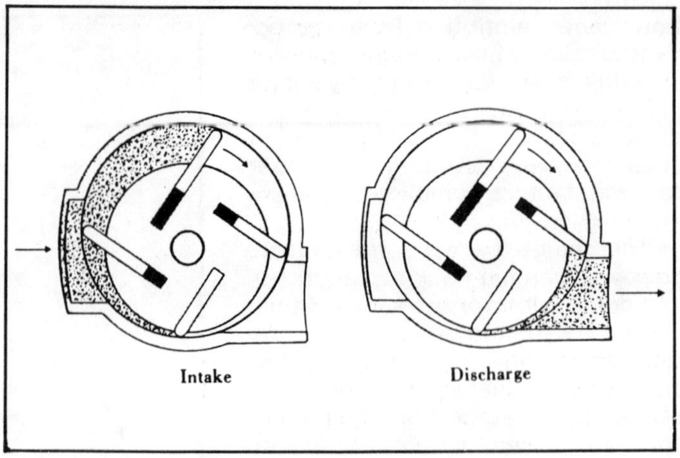

The principle on which the Judson blower operated was starkly simple. The flat vanes simply rode in slots in the rotor, and were pressed against the blower housing by centrifugal force. The angle of the vane slots, as shown here, supposedly "balanced" the weight of each vane so that they would rub against the housing with minimal force. However, constant lubrication of the vane tips was essential to blower life.

Mr. Judson is still in business in Conshohocken, these days making infrared sensors and other electronic devices primarily on a consulting basis. He no longer produces superchargers or parts for them.

THE BENDIX SUPERCHARGER

The Bendix Corporation, a major components supplier to several automotive manufacturers, has been making vane-type compressors and pumps for years—initially for auto air conditioning systems, and more recently for smog control air pumps. The prototype Bendix vane-type supercharger partially derived from this background.

At least the fact that Bendix chose the eccentric-vane blower configuration partially derives from their experience with this design. However, the fact that Bendix decided to develop a blower, and the reasons for doing so (as described in SAE paper #810006, Supercharging for Fuel Economy), should be of particular interest to any student of supercharging.

Bendix built their blower to improve overall vehicle fuel economy. Now, while several supercharger or turbocharger salesmen will try to tell you that a blower can reduce fuel consumption while it increases the power of a given engine, we know that this just isn't the case. It takes more fuel to make more power—and blowers pump it right in.

But Bendix isn't talking about getting more power out of a given engine. Their approach is to get *the same* power out of a *smaller* engine. Here's the way it works. Let's say you are a Detroit engineer and you are designing a car that weighs 2200 pounds...a typical downsized car of today. You want this vehicle to be able to accelerate from 0 to 60 mph in 15 seconds, and you calculate that a 70 horsepower engine is needed to provide this level of performance. In normally aspirated form, the engine might have to displace 1.8 liters to make this power. However, you could get the same horsepower—and the same level of performance—from a 1.3 liter engine supercharged at 6 pounds of boost.

The smaller, supercharged engine might get a little bit better gas mileage than the larger unblown engine when both are producing 70 hp, just because the smaller engine will have slightly less internal friction. But it still takes basically the same amount of fuel to make 70 hp in either motor. The fuel saving comes when the car *isn't* accelerating, which in a typical driving cycle is about 90 per cent of the time. So by fitting the 2200 pound car with a blown 1300 cc engine instead of a naturally aspirated 1800 cc, it can cruise around with a thrifty 1300 cc motor the vast majority of the time, yet have the same power as the 1800 cc engine when the throttle is opened and the supercharger comes into boost. In other words, Bendix is saying what blower people have always known—that adding a blower to your engine is the same as increasing the displacement of the engine (or swapping in a bigger motor)...except that the blower only works on demand.

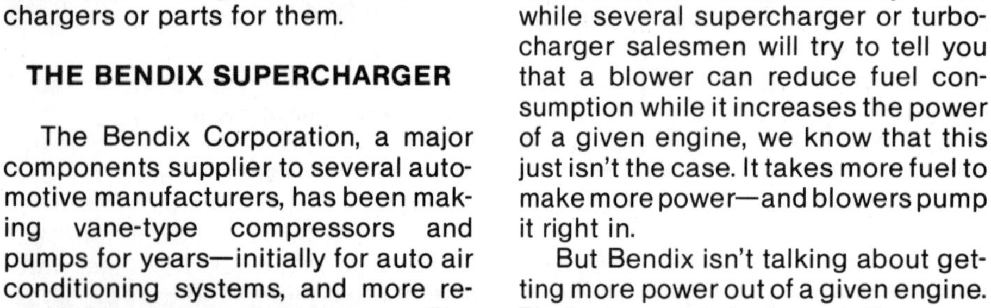

The prototype Bendix superchargers were made with an extruded aluminum housing and rotor, so that different sizes of the blower could easily be made just by cutting these components shorter of longer. As shown, the Bendix used an electrically-operated clutch drive on the blower drive pulley, similar to those on Bendix air conditioning pumps.

Well, that's not totally true. Belt-driven blowers do cause some parasitic power losses in the engine even when they are "coasting," although it is much less than when they are making boost. In either mode, some types of blowers are more parasitic than others (that is, their mechanical efficiency ratings differ).

It was the intention of the Bendix Corporation to (1) design a blower that would have minimal parasitic power (i.e., fuel) drain on the engine when not in the boost mode, and (2) would be practical and inexpensive to produce in mass quantities for installation on production-line vehicles.

First let's take a look at the design and construction of the blower itself.

The housing, or outer case, of the Bendix supercharger is an aluminum extrusion. This lowers production cost and also allows the blower to be made in different sizes (displacements) simply by cutting the housing and rotor components to different lengths. Initial test blowers were made in two displacements, 50 cubic inches and 72 cubic inches.

Inside this housing, as in the Shorrock and the Judson, a cylindrical "rotor" revolves on an off-center axis. Through slots in this rotor protrude three vanes anchored to a concentric shaft. To eliminate the internal oiling problem of other eccentric-vane blowers, however, the Bendix uses special carbon seals between the vanes and the slits in the eccentric rotor (which provide both air sealing and friction lubrication), and permanently lubricated shaft bearings. Thus the blower requires no lubrication at all. The only drawback to this design is that fuel will attack the carbon vane seals, so the Bendix blower (unlike the Shorrock or Judson) cannot be used in a draw-through configuration, but must blow into a pressurized carburetor or other fuel system (such as direct-port timed fuel injection) instead.

A big difference between other eccentric-vane superchargers and the Bendix, however, is that the prototype Bendix does not have internal compression, but is designed to be a backflow positive displacement compressor instead (as a Roots blower is). The difference is in the outer case which, instead of having intake and outlet port openings cut directly into the cylindrical wall, is extruded with an inlet cavity along one side, and an outlet cavity along the other (see illustration), spaced about 120 degrees apart. These cavities, or "bulges," along the sides of the case connect to respective inlet and outlet orifices in the blower end plate. Thus the spinning rotor and vanes simply draw air in one side, pump it out the other, and create boost in the manifold (approximately 5.5 to 6 psi on most small motors).

The reason for this design is to reduce the power drain of the blower when it is not making boost. An internal compression supercharger requires input energy to compress air on each revolution, regardless of whether it is making boost or not. The amount of air it compresses per revolution will lessen as the throttle is closed (on draw-through designs), but it will always be doing some work. On a blow-through design (like the Bendix), an internal compression blower would be drawing in and compressing a full charge of air on each revolution, and might also cause some serious pulsing of the airflow through the carburetor. A backflow (or external compression) blower, however, will begin to compress air only when the blower pumping rate exceeds the engine's intake (vacuum) rate—generally at wide open throttle under load. Thus during idle, part-throttle, or cruise conditions the backflow compressor will "rob" minimal horsepower (and fuel) from the engine.

The drawback to backflow compressors is that they become very inefficient at high boost levels. However, the Bendix engineers calculated that their blower would maintain good adiabatic efficiency up to pressure ratios of 1.6:1 (approximately 8.8 pounds of boost), while maintaining the high volumetric and mechanical efficiencies of the eccentric-vane blower design. Since this blower is designed expressly for moderate boost on factory stock engines, the reasoning is sound. And, since this vane blower is a positive displacement compressor, it does not exhibit the poor transient response (low boost at low and mid-range rpm) and "lag" characteristics of a centrifugal blower or turbo.

However, to further increase the mechanical efficiency of the Bendix blower, the designers first installed an electrically-operated clutch on the V-belt drive pulley (very similar to an air conditioner pump clutch) so that the blower could be totally disengaged during non-boost operation, as well as to guard against over-revving the blower (it is rated at 6000 rpm maximum, and in test models it was driven at 1.2:1 times crank speed). However,

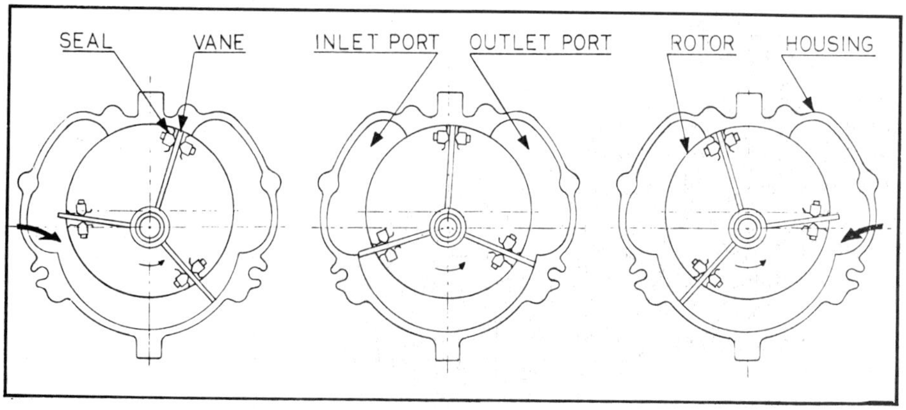

The Bendix blower actually used on test vehicles did not compress air in the housing, as other vane blowers do, but acted as a positive displacement pump. The large inlet port and outlet port cavities, which extended down each side of the case, determined this operation.

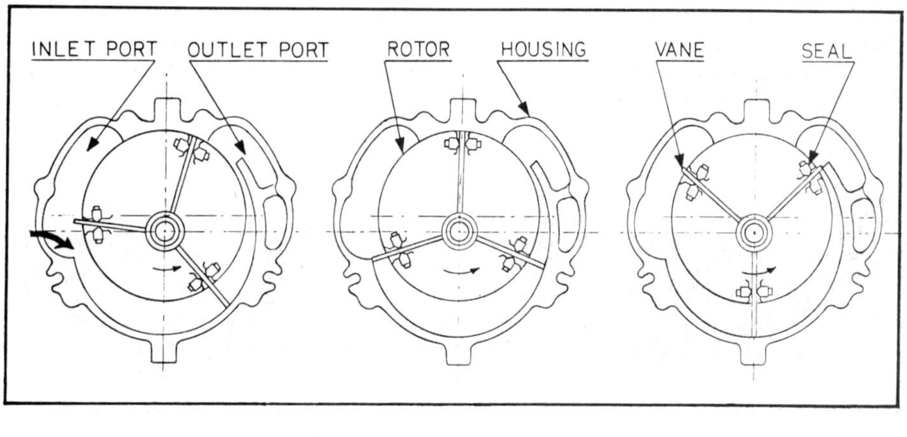

Simply by extending the internal surface of the blower housing by adding a lip to the outlet port cavity, as shown, the Bendix could be converted to an internal-compression supercharger. This would increase the volumetric and adiabatic efficiencies of the blower, but would require more horsepower to turn it—that is, lower its mechanical efficiency.

simply engaging the drive clutch when boost was needed caused too much of a jolt on the system and too sudden of a transition in engine power (remember we're talking about a prototype factory-installed system, not a high performance bolt-on). Consequently the Bendix engineers also added a "recirculation control valve" between the inlet and outlet ports of the blower, which is operated by a pressure diaphragm that reads the pressure differential above and below the carburetor. During high manifold vacuum normally-aspirated driving conditions, the recirculation valve connecting the blower ports opens to allow the blower outlet pressure to recirculate back into the intake. Under low manifold vacuum conditions (wide open throttle, heavy load), the valve closes to direct all or partial boost to the carburetor.

Obviously this system will cause some heating of the intake charge, which will lower the blower's adiabatic efficiency. Prototype installations made use of both the clutch cut-out and the recirculation valve in various combinations to control boost. Prototypes tested by various automotive publications, such as a 2.3-liter Mustang, a 2.2-liter Dodge Charger, a VW Scirocco and a Chevy Citation seemed to respond very effectively, with smooth and quiet blower operation. However, as we said at the beginning of this chapter, the Bendix supercharger reported here was strictly a prototype, and installations were submitted to various manufacturers in hopes of securing an O.E.M. contract. As of this writing, the Bendix was not being produced in the United States, although we were told that it was being built under contract in Japan.

Although several Bendix blowers were installed on prototype factory test vehicles, such as this 4-cylinder, fuel injected Chevrolet Citation, none were put on the market. In this photo the "recirculation" tube connecting the inlet and outlet ports (arrow) is visible, which allows the blower to "freewheel" when boost isn't required. Also note that the blower is driven with a single V-belt.

THE HOLLEY/WANKEL

We mentioned earlier that the Wankel Research company of Germany had developed a vane-type supercharger used very successfully on a record-setting NSU motorcycle years ago. Also, the Wankel-based rotary engine, as used in Mazda RX-7's, though very different from their vane-type blower, is still very similar in principle to the eccentric-vane superchargers we have been discussing in this chapter. The main difference between a Mazda rotary engine and a Bendix, Shorrock, or Judson blower is that, rather than having vanes slide in and out of a cylindrical rotor, the engine uses a "fixed vane" rotor, if you will, of a more or less triangular shape. This rotor, instead of simply spinning on an off-center axis, spins in an elliptical pattern about a planetary gear. The outer casing, rather than being cylindrical, is shaped like a fat figure eight, and the rotor tips maintain constant near-contact with the case wall. As we stated previously, it would be possible to convert a rotary engine to work as a supercharger, but no one has done so to our knowledge.

However, the Wankel Research vane-type supercharger which the Holley Special Products Division of Colt Industries has contracted to develop for possible new car or after-market use in the U.S. is a completely different unit than the Wankel rotary engine. It is also significantly different from other eccentric-vane superchargers discussed in this chapter.

The Holley supercharger uses a cast aluminum case which is machined with one large cylindrical surface in the center and a smaller cylindrical "bulge" above, and intersecting, the larger one. An outlet port is cast into the case wall near the intersection of the two case cavities. The inlet port is located in the center of the blower end plate, where incoming air enters a chamber inside a cylindrical

The new prototype Holley supercharger, based on a Wankel design, is actually a single-vane blower. Current test models are being made in 800 and 1400 cc displacements, with a cast aluminum housing and a clutch-hub, dual V-belt drive. This blower draws air in through a passage in the middle of the rotor, with an inlet port located at the back of the housing (arrow). The outlet ports can be seen clearly in this photo.

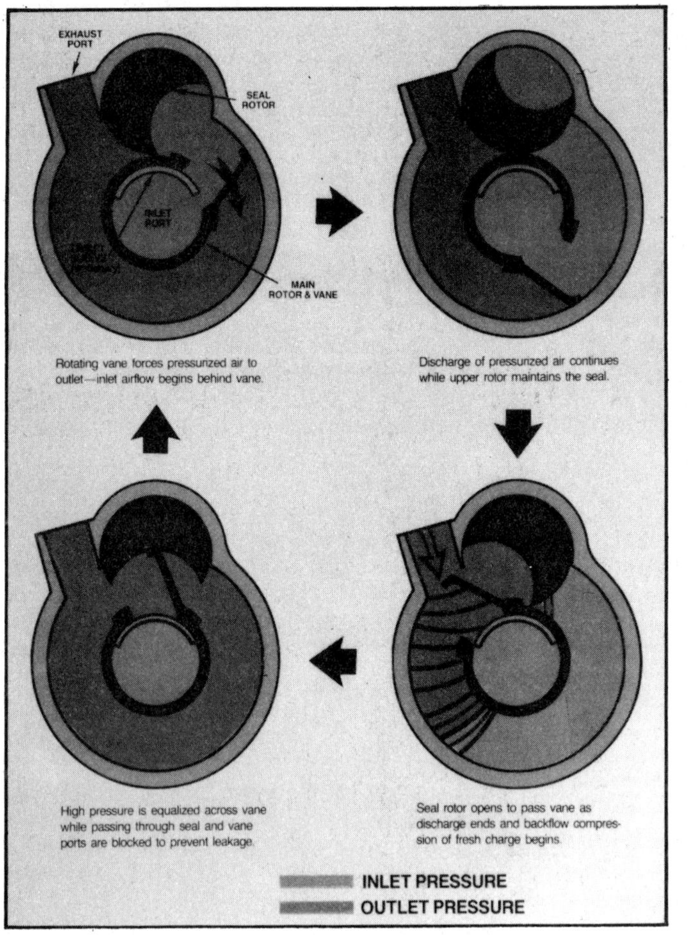

The only way to adequately describe the operation of the Holley blower is to show a diagram of how it works. It's really quite a simple design.

Only a few Holley blowers have been installed on test cars. This 800cc model has been fitted to a 1600cc Ford in a front-wheel-drive vehicle, and uses a single V-belt to run the blower. Also note the recirculation tube installed between the blower inlet and outlet tubes, with a diaphragm to operate the valve.

rotor at the center of the blower.

The rotor at the center of the blower has an opening along one side through which incoming air is ported into the blower case. A single, fixed vane is attached to this rotor, at one edge of the inlet opening, and it maintains near contact with the case wall, around the circumference of the larger "cylinder."

As the central rotor turns in the blower housing, pulling intake air in behind the vane, and pushing outlet air ahead of it, a second rotor directly above it turns in the opposite direction. The upper rotor maintains a seal against the lower rotor as the two turn against each other, so that outlet air is directed out the port. However, a crescent cut in the side of the upper rotor is timed (by bevel gears in the front blower case between the two rotors—as in a Roots blower) so that it "opens" to allow the vane to pass by while still maintaining an air seal. A "canopy" over the top of the intake port closes this inlet as the opening in the center rotor passes over it. It sounds complicated, but the accompanying diagrams should clarify how it works.

Here is how Holley describes their blower: "The Holley Supercharger is of simple cylindrical geometry and all-metal construction, having two balanced, counter-rotating elements. The main rotor is hollow as is the attached vane, thus providing the inlet air passage. A fixed, non-rotating sleeve inside the main rotor serves to valve the inlet flow. The smaller, second element acts as a rotating seal between the main rotor and housing wall. The two are synchronized by permanently lubricated gears located outside the pumping chamber in an integral gearcase. Both rotors are supported on lightly loaded anti-friction bearings and, with the exception of the bearing seals and gear teeth, there are no areas of rubbing contact or wear. Discharge is through a peripheral port in the housing wall." Is that any clearer?

The advantages of this blower design are that it maintains the close tolerances, and therefore the high volumetric efficiency, of other vane-type blowers without complicated, frictional, internal components which require lubrication and are subject to wear. More importantly, this vane blower is counterbalanced by the opposite-rotating internal elements, so it does not experience an out-of-balance shake as do other types with eccentric rotors.

Thus the Holley/Wankel vane-type blower is capable of high rotational speeds, and therefore high output, without vibration. It has very low frictional losses, virtually no wear of internal components, and therefore no need for maintenance. Early prototypes of this blower being tested at the time this book was being compiled were fitted with an electric drive clutch similar to that on the Bendix supercharger (and on A/C compressors), these being grooved for two V-belts. The one installation shown was of a blow-through configuration; we do not know if this blower could be used in a draw-through arrangement or not. Two sizes of prototypes were being tested, one of 800 cc's and another of 1400 cc's displacement per revolution. They are identical except for the length of the case and rotors.

Beyond this, we have no further information on the Holley supercharger at this time. We stress again that this blower is strictly in the development and testing stage at Holley. It will probably not appear on the market unless Holley secures a contract with a major manufacturer to install it on new cars. At present it is not available to the public.

CHAPTER 7 SUPERCHARGING ON THE STREET

Many of today's new, smaller superchargers which produce between 5 and 9 pounds of boost are designed to bolt directly onto our current low compression "smog motors" with virtually no internal modifications. The installation of a B&M blower on this new Camaro is a typical example.

BUILDING A BLOWN STREET MOTOR

There are two ways to approach the construction of a street supercharged engine. If you are going to be adding a modest boost blower kit (one which will produce a maximum of 5 to 8 pounds of boost) to a stock, late model, low-compression "smog" motor, you probably will not need to make any significant changes to the engine before bolting the blower in place. Superchargers such as the B & M and the Penco for V-8's, and the Max 25 or the smaller Magna Chargers for 4-cylinders or V-6's, are excellent for this sort of application. *Rather than* pulling the engine out, tearing it apart, and changing pistons, rings, bearings, cam, valvetrain, heads, manifolds, carburetor, ignition, and so on, you can buy a supercharger kit, *bolt it on* to the engine in an afternoon, and drive away with a strong, usable 30-40% horsepower increase without doing much of anything else to the car.

Contrasted to traditional unblown performance modifications, a supercharger making in the neighborhood of six pounds of boost will greatly increase both horsepower and torque in the 2500 to 4500 rpm range, where you can really feel it in a street-driven vehicle, and will continue to produce power right on up the rpm scale. On the other hand, an engine which has been fitted with higher compression pistons and a longer duration cam to achieve the same horsepower increase as the supercharged, but otherwise stock, engine will not only have less power in the lower rpm ranges, but will be rougher running and

The new dual Magna Charger is advertised to give up to 12 pounds of boost on 91 octane gasoline on an otherwise stock 8- to 8.5:1 engine without even the need for water injection. Jerry does recommend his own blower camshaft as one addition, but installations on stock late-model V-8's, such as this '82 Camaro, seem to give plenty of street horsepower with no modifications other than the blower kit.

On the other hand, if you're intention for bolting a big blower on your motor is to make some real horespower, then you had better build the motor the way the pros do. Rodney Martin, of the Super Shops, chose to use two 650 double-pumper Holleys instead of injectors, a 1:1 blower drive ratio, copper head gaskets with stainless O-rings, 9.0:1 compression, and a well-designed cooling system on this 650 hp, 355 smallblock in his '40 Chevy pickup. Rodney uses this "eye-opener" as daily transportation to and from work.

probably less efficient at street and normal highway speeds.

Since most of our passenger car engines made since 1972 come with compression ratios in the 7.8:1 to 8.8:1 range, the addition of a supercharger is a very practical, relatively quick and simple way to achieve a healthy performance increase. As long as you keep blower boost below approximately 9 psi, and you don't rev the engine above 5500-6000 rpm, you shouldn't really have to make any internal modifications to the engine in order to accept the blower. If anything, you might want to install slightly stiffer valve springs to help keep the intake valves on their seats, and you might want to switch to a good street blower cam—though most of today's "smog" cams will work well in a mild blower application as long as they aren't retarded severely to help burn hydrocarbons in the exhaust manifolds (if so, you can simply advance the cam timing).

Obviously the above statement—that you can bolt a 6-9 psi blower system onto a stock "smog" motor without making any other modifications—requires some qualification. First, of course, we're talking about an engine that's in good condition to start with. Secondly, as we said, we're not talking about an engine that is going to be drag raced every weekend in a four-speed car at 8000 rpm shift points. But most importantly, whether or not the mild street blower will be compatible with a stock low-compression engine will depend on the point at which the combination will begin to ping or detonate. This point will depend on the design of the combustion chamber in the particular engine, the efficiency of the blower, your driving habits, the weight of the car, and the quality of gasoline available, among other things.

My point is that the life expectancy of the parts in *any* street-driven supercharged engine these days depends primarily on whether you can safeguard those parts from detonation. You probably won't be able to run your supercharged engine on unleaded gasoline. You will probably have to richen the carburetor jetting somewhat, and you might even have to retard the ignition timing slightly. Perhaps you will have to add a water injector. But you certainly do *not* want to rattle the pistons in a stock engine. But with good gasoline and proper carburetor jetting, you will find that six to eight pounds of boost is very livable in a stock motor. Further, remember that in a street-driven supercharged vehicle, the blower actually gets into boost very seldom, and then usually only for short bursts. The rest of the time the engine operates under no greater loads than normal.

If you have any doubts about bolting a mild blower onto a stock low-compression engine, my personal advice would be to go ahead and install the blower on the car, being sure to add a boost gauge so that you know exactly how much the unit is putting out. If you experience knocking in the motor when the blower is under boost (and you are already using the best gasoline available), start by richening the secondaries of the carburetor, make sure that the distributor isn't overly advanced, and then consider water injection. If the boost is kept in the 6-8 pound range, I seriously doubt that you will have any problems. If for some reason the engine *can't* handle this boost, and you blow a head gasket, or burn a valve, or even break a piston land or ring, *then* you can pull the motor out and go through it. If you are the type of person who would rather be safe than sorry, it certainly wouldn't hurt to blueprint any engine which will be driven in any sort of performance mode. But with mild blower boost (below 9 psi), and low static compression (below 9:1), the instances in which production parts will fail are the exception, by far, rather than a common occurrence.

If you plan to bolt a 6-71 with a pair of four-barrels on your small block Chevy, and you fully intend to get your foot into the throttle now and then, you better think about upgrading some of the components below that blower. Again, many of the same conditions apply that we mentioned above—if you can keep your big blown motor out of detonation, you're not going to have to worry a whole lot about parts damage. Likewise, if you are not going to abuse the engine, over-rev it, or race it every week, you don't need parts like a steel crank, four-bolt mains, O-ringed heads, and so on. Many customers have bolted 4-71's, Magna Chargers, Paxtons, or even 6-71's onto relatively stock low-compression engines without any problems at all. But once you start making boost in the 10-15 pound range, or buzz the motor to six grand or more, you are certainly in marginal territory with factory components—especially in terms of pistons, rings, and head gaskets.

Here's the bottom line: If you are adding a supercharger to your motor to make it a high performance engine, and you plan to drive it in a high performance fashion, then you better *build* the engine to high performance specifications. *Any* high performance engine, supercharged or not, should be blueprinted—this includes checking and fitting all clearances, matching ports, dialing in cam timing, balancing the rotating and reciprocating assemblies, and so on. It is not within the scope of this book to fully describe engine blueprinting techniques, but the procedures for a supercharged engine are basically the same as those for any high performance engine. In the following sections we will discuss peculiarities in parts selection or assembly procedures unique to a blown engine, but there aren't many, actually, that differ from any performance engine blueprint.

It isn't practical to try to run more than 15 pounds of boost on any street engine. In the first place, Roots-type blowers aren't very efficient above this level, and in the second place it would be very difficult to find any type of gasoline that would be compatible with such boost on the street. Consequently, few of the problems associated with superchargers in racing applications apply to street engines. You certainly aren't going to be blowing out cylinder walls or breaking crankshafts because of extreme cylinder pressures. Hopefully your blower will be operating in a fairly good efficiency range so that it won't be heating the intake charge excessively, and therefore causing extreme temperatures in the cylinders (as a big GMC certainly would, especially on gasoline, if it was cranked up to 20 or 30 pounds of boost).

The only real problem to watch out for in a supercharged street engine—primarily those set up to make over ten pounds of boost—is that the blower *will* inevitably cause some increase in intake air temperature, which will cause both the cylinder temperature and exhaust temperature to be higher than they would in normally-aspirated engines at similar power levels. However, in street supercharged motors this extra heat should be easily handled by a good cooling system, high quality pistons and rings, good valves and a proper valve job, and good head gaskets. Remember that even a high performance street blown motor still sees the full stress of maximum boost very rarely—as opposed to a race car, which is built to operate under constant boost. The rest of the time the engine will be running on a much lower effective compression ratio than a comparable normally-aspirated high performance engine. Also keep in mind that, when a blown motor is under boost, the denser air charge forced into the cylinder "cushions" the piston/rod/crank assembly during the cycle. A normally-aspirated, high-compression engine actually sees much higher *peak* cylinder pressures than a comparable blown engine, and this can cause greater stress on parts, as well as promote detonation.

Before we look at the individual parts of the engine, and discuss how they might be reworked for a high performance blower application, let me introduce a currently-available factory-built 350 Chevy small block which would serve as an excellent prototype

A good starting point for a high-horsepower street supercharged engine would be something like Chevrolet's LS-6 or LS-7 454 big blocks, which are still available from the factory as shown. This one was ordered by Art Chrisman to build for a customer with a 6-71 Mooneyham blower and dual four barrels to install in a street rod. Art will tear the new motor apart, blueprint it, add forged 7.5:1 pistons and steel rings, and a complete Crane cam and valvetrain kit before installing the blower.

Another excellent choice for a stout blower motor is the Chevrolet heavy duty marine 350 small block, shown here with a 6-71 Jimmy and Isky drive installed. This engine comes with 4-bolt mains, a steel crank, good rods, and 8.8:1 forged pistons—it is perfect for a blower as-is. If Chevrolet re-introduces the LT-1 small block (as rumored), it would also be a good base for a blower, although lower compression pistons may have to be installed.

for most any high performance street engine build-up. It is called a Heavy Duty Marine 350, and it is available only through volume marine engine dealers such as Mercruiser, Volvo, or Gale Banks Engineering.

This engine comes with a heavy wall (but not siamesed cylinder) 4-bolt main block, a forged steel crank, LT-1-type "pink" rods, and flat-top forged pistons rated at 8.8:1 compression. The heads have 1.94-inch intakes and 1.50-inch exhausts, large chambers, induction-hardened exhaust valve seats, and good-flowing "pre smog" ports. Since it is a marine engine, it also comes with brass freeze plugs, stainless steel shim head gaskets, and a brass impeller water pump. Further attractions include the heavy duty truck roller timing chain, a windage tray, a good-flowing cast iron Quadrajet intake manifold, and even finned aluminum Corvette valve covers. The camshaft is similar to an L-82 hydraulic (approximately 221° duration at .050-inch lift), which would work nicely with a blower. The only part on this motor not compatible with a high performance application are the exhaust valve rotator-type spring retainers, which can easily be replaced with standard retainers and springs after the springs are shimmed approximately .120-inch.

This engine is actually *overbuilt* for most street supercharging applications. The average street blown motor doesn't really need four-bolt mains and a forged steel crank (especially non-Chevies, for which such parts may not exist). However, this combination would not be all that hard to duplicate in a small block Chevy, if you started with a good pickup truck 4-bolt main block. If you are considering building such an engine, you might consider buying one as-is. As of our publication date, they retailed for approximately $2200, complete as described above, from Gale Banks (Irwindale), California). Gale uses them, without any changes other than valve springs and retainers, for his 500 horsepower dual turbocharger set-up for boats.

BLOCK, CRANK, RODS

There's not much to say here that we haven't just stated above. If you are going to be using a moderate boost street blower, you probably do not need to tear into the short block. If, on the other hand, you plan to make over nine pounds of boost and you figure the engine might be seeing some high rpm, then you should plan on a typical performance blueprinting of the entire block and lower engine assembly.

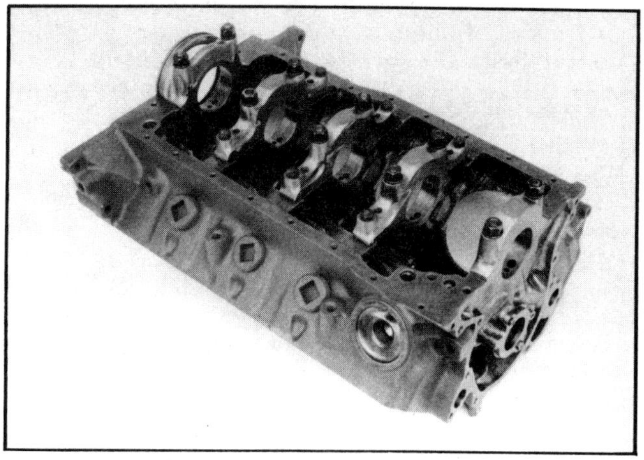

Although a four bolt main block would be preferable for any high performance engine, it is certainly not a necessity for most street blower applications. Engine parts selection depends more on the amount of boost you intend to run and the amount of horsepower the engine will make, rather than the fact it has a supercharger.

Magnaflux the crank and rods, straighten and polish the crank, chamfer the oil holes, shot peen and polish the beams on the rods, resize all the holes and align all surfaces, and so on. There's nothing special a blown motor needs that any high performance engine doesn't need in this department.

As we mentioned, a four-bolt main cap block and a forged steel crankshaft are certainly not requisites for a supercharged street engine, unless you are planning to make around fifteen pounds of boost and intend to punish the motor with it. On the other hand, four-bolt mains and a good crank are excellent insurance, and such blocks for Chevies are relatively easy to find in pickups or larger trucks in the wrecking yards. Billet 4-bolt main caps are available from specialty shops for many types of engines, but these are usually quite expensive, require custom machining of the block, and really aren't necessary for anything but an all-out racing engine.

If you are using a cast crankshaft, it would be wise to give it a surface-hardening treatment, such as Tufftriding;

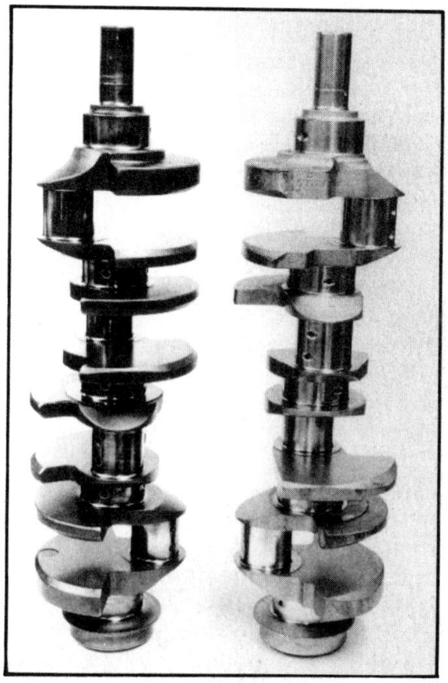

Leave the super expensive steel billet custom crankshafts (left) to the professional racers. All a street blown engine needs is a stock crank, which can be improved with chamfered oil holes, polished journals, and Tufftriding (right). A factory steel crank would be preferable, but cast cranks will work fine in most street applications.

and choose a crank with rolled journal fillets if you have a choice. Since you cannot perform any crankshaft preparation work in your own garage, all you really need to do is give your crank to a good engine shop and ask them to prepare it for the use you expect your engine to see. They should know what to do.

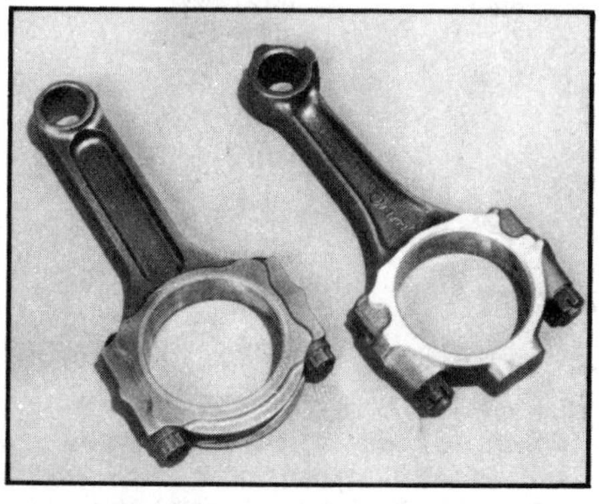

Likewise, stock connecting rods are all any street blown motor needs, especially if they have been Magnafluxed, shot-peened, polished on the beams, and fitted with SPS bolts like the Cleveland Ford rod on the right. Leave the super-trick Carrillo or Crower rods (left) for the long-distance racing engines.

Nonetheless, it seems that lots of guys who build blown street engines get a kick out of spending money. A typical case in point are high-dollar connecting rods. In visiting countless engine shops over the years, I have seen far more sets of the "super trick" forged or billet steel connecting rods

Only on very high boost street engines will O-ringing of the block be necessary. This NASCAR Ford engine has been fitted with flat Cooper rings around each cylinder bore, and the gaskets have been trimmed away accordingly. Also note the use of head studs to insure even and positive torquing of the head.

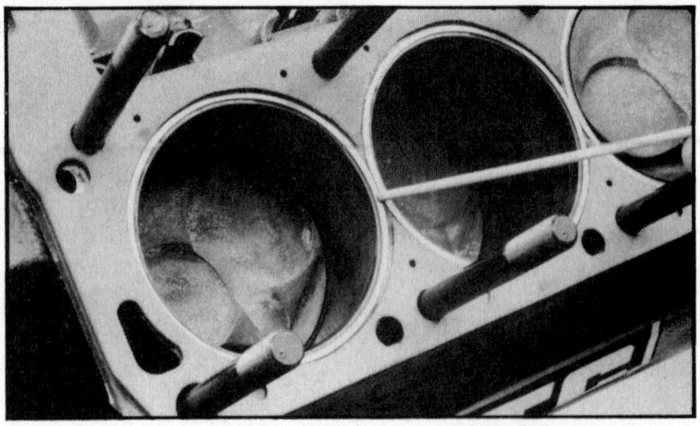

being installed in street engines than I have in actual race motors. I'm certainly not saying that certain types of racers don't use—and need—the good rods...but I am saying that street engines surely don't need them. How often have you seen a supercharged street engine throw a rod? Or break a crank? Personally, I've never even *heard* of such a situation. Broken pistons or rings, yes, or blown head gaskets, or even valve problems, but not bottom end failures on a streeter.

Your main insurance on a street engine is to check (Magnaflux, Zyglo, etc.) the stock components for cracks or fatigue as part of the blueprint. As far as the connecting rods are concerned, the weakest link is usually the bolts. Always replace used rod bolts when rebuilding a performance engine—and it is wise, and relatively inexpensive, to use a stronger SPS-type bolt as a replacement. On engines such as the 289-302 small block Ford which came with marginal 5/16-inch diameter rod bolts, it would be advisable to drill the rods for larger bolts (such as 11/32-inch Chevy bolts), or switch to much stronger High-Performance 289 rods with 3/8-inch bolts.

Bottom end clearances in a supercharged engine should not differ from those in any high performance application. All engine builders that we interviewed for information in this section of the book strongly recommended TRW Clevite-77 (CL-77) main and rod bearings for a supercharged engine. These readily-available bearings feature a tri-metal surface, with a lead-tin babbitt overlay on the top, which is softer than that of standard bimetal bearings, yet still has a higher load-carrying capability.

Engine builder Mike Kuhl of Santa Ana, California, who has extensive experience building blown engines of all types, including lots of river-running boats, pointed out that one critical block preparation for any blown engine is to resurface both the cylinder block decks as well as the heads to make sure that all of these surfaces are flat so that head gaskets will seal properly. Kuhl, who works primarily with high-boost motors, also recommends O-ringing the block on most engines, but especially big block Chevies. "I wouldn't run a 454 without O-rings—at any boost," he states flatly (though other builders of milder-boost street motors might disagree).

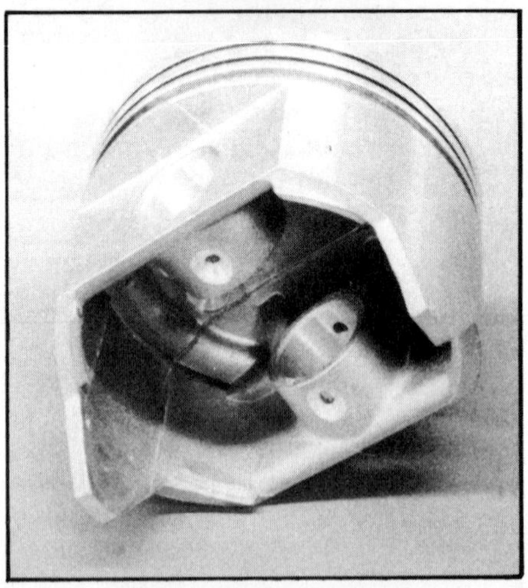

If your engine has a set of pistons in it that look like this, do *not* bolt a blower on top. These are inexpensive, cast, pop-up pistons for a small block Ford which give 10.5:1 compression—far too much for a blower. Also, do not have the domes milled off such pistons to lower the compression, since the remaining top surface would be too thin and too weak to sustain even moderate blower pressures.

Kuhl also sells a special oil cooler for his high performance blown boat engines, as well as a special pump. On any street blown motor, you want to make sure that lubrication is efficient and adequate—high performance oil system parts and techniques vary widely from one brand or type of engine to another (once again, follow typical performance engine procedures). However, even moderate-boost blown street motors could benefit from a larger-capacity oil pan, since the greater volume of circulating oil will help to cool the engine (and the cooler oil will live longer, too). However, this is certainly not a necessity on most street engines.

PISTONS AND RINGS

This is the primary area in which the construction of a street supercharged engine differs from that of other high performance engines. Let's start with some basics.

Number one: you cannot run high compression pistons with a blower. For all practical purposes, the static compression ratio in a supercharged engine must be *below* 9 to 1. How low should you go? Unfortunately there just isn't any rule or formula for "balancing" static compression ratio with blower boost. For street blown motors, however, nearly all engine builders set the static compression between 8:1 and 8.8:1. Then you can "dial in" the blower to provide the most performance possible up to the point of detonation. For the bigger blowers, such as 6-71's, you will probably want a compression ratio towards the bottom of this range (8:1), while the smaller blowers (B&M, Penco, etc.) will work well closer to 9:1. Of course there are *lots* of variables involved here, including engine size, combustion chamber size and shape, and so on.

Number two: you can't tell *what* the compression ratio in the engine is unless you measure it. Factory engines can vary a point or more above or below the specified compression (in fact, factory engines can often vary considerably in CR from cylinder to cylinder—that's just one reason why blueprinting is necessary). Even aftermarket pistons which are rated at a certain compression ratio must be matched with the specific components in your engine. Minor differences in connecting rod lengths, piston pin heights, dome heights, valve head heights, valve seat depth, and combustion chamber core shift can all add up to alter the actual reading considerably. Furthermore, "listed" compression ratios for aftermarket pistons are stated for a specific volume combustion chamber and a certain thickness of head gasket. Combustion chamber sizes have varied over the years on most types of engines; and, if you are using older heads, how many times have they been surfaced in the past, and how much?

Figuring the exact static compression ratio in an engine involves a lot of measuring and a lot of arithmetic (although it *is* simple arithmetic), but it is the only way to be sure how much compression the engine really has. Obviously, we are talking about a serious engine build-up—a high performance engine that will see some healthy boost and will therefore require a tear down, parts upgrading, and blueprinting. If you are going to bolt a mild boost blower onto a relatively stock engine—that is, if you aren't planning to change the pistons anyway—there's little point in pulling the heads off to check the compression. This would only be necessary if your engine were encountering a severe detonation problem that could not be eliminated by other means.

The above statement is important, so let's phrase it another way. If you are starting with a late-model, low-compression engine (in the 7.8 to 8.8 to 1 compression range) in good running condition, and you want to add a bolt-on, moderate boost (5 to 8 psi range) blower kit to improve performance, there is no need to pull the engine apart either to change the pistons or to check the actual static compression ratio. Add the blower, dial in the system (jet the carburetor, set the timing, add water injection if absolutely necessary), and figure that everything is going to work fine. If the engine does experience severe detonation under boost, even though you have taken all of the "external" steps to eliminate it, *then* consider pulling the heads, measuring the actual compression ratio, and possibly changing to pistons of a lower ratio, of higher quality material, or both.

In the rare case that you might want to add a moderate-boost street blower to a stock engine old enough to have a high compression ratio (say 10:1), you might *still* want to try the blower first (remember we're talking low blower boost here) with good gasoline, very conservative ignition timing, and possibly water injection, to see how much detonation you encounter. If it rattles, change the pistons. But at 5 or 6 pounds of boost you might be able to get away with it. On the other hand, any

This was the sorry result of one instance of trying to lower the compression ratio of an engine for a blower installation by milling the tops off of the pistons. Even though the pistons were forged, the milling operation left too little piston crown above the top ring land, and the lands broke off under blower boost complicated by detonation.

high compression factory engine is going to be at least ten years old these days, and unless it's been recently rebuilt, it is going to be tired. If an engine has worn pistons, rings, valves, or other components, adding a blower is only going to aggravate these problems. This is important, so let's say it again: if your engine has any blowby, leaky rings or valves or guides, sloppy pistons, or marginal head gasket seal, adding a blower is certainly going to worsen the situation. The point is, don't add a blower to a worn engine. Furthermore, adding a blower to a relatively fresh factory engine, or a standard rebuild, might (in isolated cases) immediately point out the flaws which sometimes creep past assembly line employees. If so, a blueprint is in order.

Now, if you are going to be adding a healthy-boost blower system to the engine, if you intend to be making some exorbitant horsepower—and using it, if you think your combination may encounter some detonation or cooling problems, or if you just prefer to be on the safe side in any case, then you should consider installing a set of forged pistons and good rings in the engine.

There are two ways to go on forged blower pistons. One is to order a set of custom-made forgings tailored exactly to your needs by one of the several good high performance piston manufacturers such as Ross, Venolia, Forgetrue, BRC, Diamond, JE, and so on. In the case of less popular engines, this may be your only choice in a low-compression forging. There are a few drawbacks, however. Naturally, such pistons are the most expensive. They must be custom-ordered from the manufacturer, and you must wait for them to be made. If it happens to be racing season (and it always seems to be, somewhere), the length of time it takes to fill your order will usually be proportional to the number of racers who need pistons (and they all do, every week) who have that manufacturer's decal on the side of their car. True forged aluminum pistons also expand more than cast aluminum pistons, so they must be installed with wider piston-to-cylinder wall clearances (.008-.009-inch clearance for forged, .002-.003-inch for cast). Consequently forged pistons will audibly rattle in the cylinders when the engine is first started cold, and you *must* allow the pistons to warm and expand before you go racing or drive down the road. This can be an annoyance in a street engine; but if you install custom forged pistons, you must allow ample engine

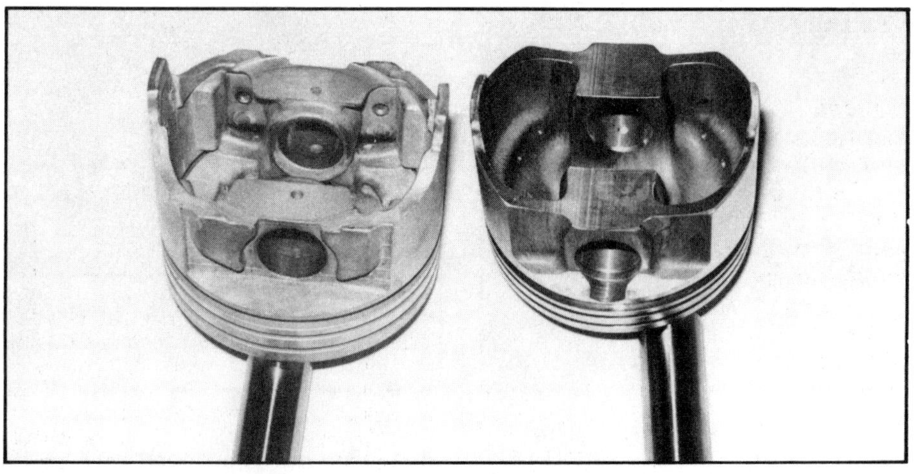

Standard cast pistons (left) will work with mild blowers in the 5 to 7 pound boost range. For any blower application, however, and certainly for those making over 10 psi, quality forged pistons (right) are recommended.

Any of the major performance piston manufacturers can make custom forged blower pistons to most any specifications you dictate. This is a 7.0:1 forging machined by Venolia to fit a blown 229-inch Chevy V-6. Note the deep dish in the top, and the considerable distance between the top and the first ring groove.

warm-up every time you start the car.

Ultimately, however, such high performance forged pistons are the best you can get. They are the only type that can be made exactly to your order. And they are an absolute necessity in an all-out high performance supercharged motor that will see lots of high boost (over 12 pounds) and hard use.

The other alternative, and the one that is in most cases more practical for a street blown engine, is to use a replacement-type forged piston from TRW or a comparable manufacturer. The advantages are that such pistons are readily available across the country, they are less expensive than custom forgings, and they are made in such a way that they expand at a rate similar to cast pistons, and can therefore be installed at clearances similar to those for cast pistons (.0025-.003-inch in most street supercharged applications). The drawback is that such pistons, in low compression ratings suitable for blower applications, are made only for Chevrolets and a few smaller engines (i.e., V-6's and fours). TRW currently offers "Powerforged" blower piston L2441F for the 350 Chevy, which has a slight dish on the top and gives 8.4:1 compression with a 76 cc chamber head, and L2453F for the 454 big block which gives 8.36:1 compression with 106.9 cc open chamber heads. If you're building a Chevy, these are excellent choices for street blown motors.

TRW Powerforged pistons are available in higher compression ratios for many popular engines. However neither these high compression pistons, nor custom forged pop-up pistons, should ever be milled down to make "blower" pistons out of them. The practice of trimming the dome of a piston in a mill or a lathe to adjust the compression ratio slightly is common, and acceptable, for normally-aspirated engines. However, this should not be done on pistons for a blown engine for two reasons. First, milling the dome makes the top of the piston thinner—and blower pistons need a *thicker* top, if anything, to withstand the increased cylinder pressure and heat produced by the blower. In most cases, milling the dome of a high performance piston enough to lower the compression adequately for blower use would severely weaken the piston. Second, milling the top of the piston lower, particularly on low-dome or flat-top designs, will reduce the distance between the piston top and the first compression ring groove. The "shoulder" of the piston top above the first ring land is an area of critical stress in the piston. If the thickness of the piston top above this groove is marginal, the edges of the top are prone to cracking, burning, or even breaking off when cylinder pressures and temperatures are increased. On a blower piston, the first ring groove should be at least approximately .250-inch down from the top.

As far as piston rings are concerned, the rule of thumb for supercharged engines is—use the best. Actually, there are three major considerations for the choice of rings in a blown motor: they must seal the cylinder against higher pressure; they must withstand and dissipate greater heat; and they must have a high tensile strength so that they will not break under heavy load or detonation.

Typical standard production or "service" rings are made of cast iron. Their major attribute is that they seat quickly, and provide fine sealing in standard engines. However, cast iron is brittle rather than ductile, and it has a melting point of 2000° F. Chrome plated iron rings are better, having a melting point of 3212° F, but they are still brittle and they are much more difficult to seat in the cylinders. A popular choice for performance engines today

An excellent street blower piston for small block Chevies is the TRW "Turbo" Powerforged L2441F, which gives 8.4:1 compression and has an extra thick crown to withstand boost and higher temperatures.

Speed Pro also offers good over-the-counter pistons made from TRW Powerforged blanks. Chevy piston 7064-P (left) is a flat top which gives 9:1 compression with a 76cc head, and would work well with low-boost (5-7 psi) street blowers. Part no. 7058-P (right) is their dished "turbo" piston with a thick crown, which gives 8.4:1 compression with 64cc heads, or 7.7:1 with 76cc chambers.

are "moly" rings, which are nodular iron rings sprayed with a molybdenum filling on the face. Molybdenum has a melting point of 4750° F, the nodular iron body is ductile rather than brittle, and they seat easily like cast iron rings. "Double moly" ring sets include moly-filled top and second compression rings, usually with a chrome oil ring. These rings (available from most major manufacturers) would be a good choice for most street supercharged engines.

On the other hand, TRW has recently introduced a new high performance piston ring called a Ceramic Plus. These rings have a *very* high heat range. They also seat more easily than chrome rings, although they do require a special "plateau finish" honing of the cylinder bores (detailed instructions are included with the rings; the process can be easily accomplished by any engine machine shop). For street blower use you would want the Ceramic Plus rectangular rings, rather than the Ceramic Plus pressure balance rings which are strictly for drag racing. In fact, these are now the only rings TRW lists for their L2453F Powerforged blower pistons.

Supercharged engine builders who are used to high-boost, high-horsepower combinations, however, such as Art Chrisman and Mike Kuhl, prefer stainless steel rings for such applications. Chrisman likes a non-cast chrome or stainless top ring "to keep them from breaking" in a healthy blown motor, followed by a moly second compression ring. Kuhl states that he uses chrome or stainless steel rings in any blown motor, since "detonation will knock the moly out of moly rings." He also suggests increasing the end gap on the top ring in a blown motor slightly (about .005-inch) to allow for greater heat expansion. In any case, with *any* rings, it is critical to check and fit the end gaps of each compression ring in each cylinder during engine assembly. Most high performance rings are made slightly longer than necessary, so that you can file them to fit. Failing to do so will allow the ends of the rings to butt when they heat and expand, causing them to seize in the cylinder, or to break.

CYLINDER HEADS

Unlike pistons and rings, the cylinder heads on a supercharged engine really need little attention. The purpose of the blower is to get the fresh air and fuel in, and the exhaust out...and that's what nine tenths of performance head work is all about. As far as port size, port shape, and valve size is concerned, as long as they aren't so small that they will strangle the blower's boost, leave them the way they are.

Hardly any street supercharged engines will need O-ringed heads. However, Chevies will benefit from the new factory high performance head gasket, which has stainless steel O-rings built in.

There are some special considerations for cylinder heads on a supercharged engine, however. First of all, you want to make sure that the heads seal in the extra pressure pumped by the blower. We would highly recommend that you have the heads surfaced to make sure they are flat. For the majority of street blower applications, you will not have to O-ring the heads (or block); this should only be necessary on very high-boost installations, or possibly on engines with poor head sealing to begin with (usually because of too few head bolts surrounding each cylinder). If you feel that O-ringing might be necessary—or you become convinced after a few head gasket failures—you will have to find gaskets to fit your engine which are compatible with O-rings. A steel-faced sandwich gasket is best (steel on each side with asbestos in the middle); a steel shim gasket will work if you can carefully match the O-ring to the rib on the gasket (this is touchy); and solid copper gaskets are best left to race cars since they are prone to water leaks. Of course the cutting of the O-ring grooves should be done by an experienced machine shop.

However, few street blown cars will need to be O-ringed. An excellent compromise these days are new high performance head gaskets which have a built-in stainless steel "O-ring" seal around each cylinder. Such gaskets are currently available for big and small block Chevies, and Chevy and Buick V-6's, with other applications bound to follow shortly. These gaskets are expensive, but they are much easier to install than true O-rings. Obviously with these, or any other gaskets on a supercharged engine, a careful and sequential torquing and subsequent retorquing of the heads is advised.

The second consideration for cylinder heads on a supercharged engine is to guard against heat and detonation. While the intake and exhaust ports probably won't require enlarging or polishing, it would be helpful to clean up the combustion chambers to remove any burrs, flash, or sharp edges which could become hot spots and lead to preignition. On a high-boost motor, you might as well polish the entire chamber, which will not only help reduce detonation, but will also help lower the compression ratio. Just don't grind so much out of the heads that you weaken the chamber walls—remember that blower boost can break through combustion chamber walls just like it can break through piston tops if the surface gets too thin.

For valves and valvetrain there are only two considerations for a super-

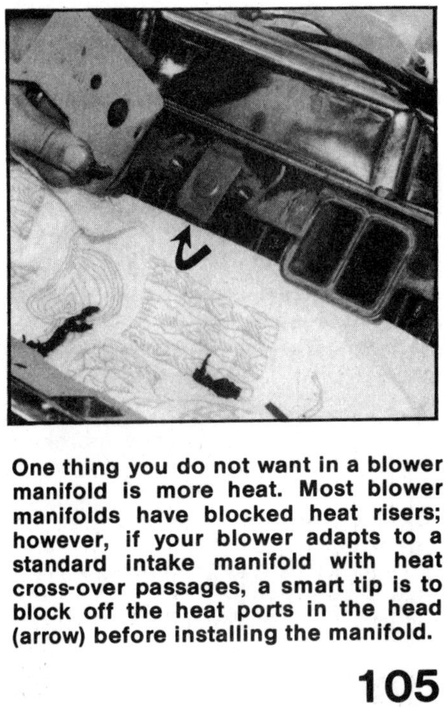

One thing you do not want in a blower manifold is more heat. Most blower manifolds have blocked heat risers; however, if your blower adapts to a standard intake manifold with heat cross-over passages, a smart tip is to block off the heat ports in the head (arrow) before installing the manifold.

charged engine which would differ from other performance engine buildups. First, since the operating temperature of the valve is controlled by its contact with the water-cooled head, and since the valve seat is the area of contact through which heat can be transferred from the valve to the head, it is imperative not to narrow valve seats (particularly on the exhaust valve) as much on a blown engine. Three-angle valve jobs are popular, and effective, for increasing airflow past the valve on normally aspirated engines. In most cases, however, cutting the valves and seats at three angles greatly reduces the width of the actual valve-seat area. If you use three-angle cut seats on the heads of your supercharged engine, be sure to keep valve seat width to *at least* .080-inch, and preferably wider (say .100-inch). Remember, the blower will do most of the job of getting air past the valves. Actually a standard valve job should be fine for a blown engine, and would be the best to insure that the valves did not burn from increased cylinder temperatures.

Second, any performance engine will usually require stiffer valve springs to guard against valve float at higher rpms and to keep the lifters on the cam with more radical camshaft profiles. But a supercharged engine will also need greater valve spring pressure on the intake valves to keep these valves on their seats, since blower pressure will be acting on the back sides of all the intake valves. Admittedly, even 15 or 20 psi isn't enough air pressure to blow an intake valve open, even with stock springs, but the addition of blower pressure to the momentum of the valvetrain components at high engine speeds compounds the spring tension requirement on the intake valves. This is critical in a supercharged engine. If an intake valve does not fully close (or "hangs up"), combustion in that cylinder will backfire into the intake manifold and into the blower, both of which are filled with pressurized, atomized air/fuel mixture—a highly volatile substance. If you have watched Top Fuel racing you know the result...and the reason why the NHRA now requires restraint straps on T/F blowers. You certainly do not want to "light off" the blower on your street machine, either. By the way, a broken intake valve spring will have the same result, so be sure to invest in a good grade of springs from a reputable manufacturer.

Of course a broken valve would be even more destructive. Most street blown motors won't require special valves, but if you are planning to make some rpm's with a stout blown streeter, experienced engine builders like Art Chrisman strongly suggest a good set of stronger stainless steel valves. Mike Kuhl also suggests P.C.-type valve stem seals for blown motors, to keep cylinder pressure from leaking up the valve stems, and he also recommends running a bit more valve stem to guide clearance on the exhaust valves.

Otherwise, head prep for a street blown motor is basically the same as for any performance engine, but without all of the time-consuming, touchy, and expensive port and polish work.

CAMSHAFTS AND VALVETRAIN

Just as it is a definite no-no to run high performance, high compression pistons in a supercharged engine, it is equally and emphatically against the rules to use a typical high performance, long duration, big overlap camshaft in a blown motor. These are by far the two most common mistakes in building street blown engines, and of the two, overcamming is the most flagrant.

Typical high performance camshaft technology has been developed to get *unpressurized* air into and out of the cylinders. This technology does not apply to supercharged engines. Actually, camming a blown engine is quite a bit easier than trying to match cam timing and profile to the various components of a normally-aspirated high performance engine (especially manifold type, size, and runner length). As we have said before, in a blown motor, the supercharger will do the work of getting the mixture into the cylinders, and much of the job of getting the exhaust out.

Even if you retain the stock cam in your street blown engine, it would be smart (and simple) to install a set of higher tension valve springs, since blower boost tends to hold the intake valves off their seats, and a valve which hangs open or floats at high rpm will cause a backfire into the blower. Shown are typical street performance springs with flat-wound dampers: Crane on the left, Chevy Z-28 off road type on the right. Always use steel retainers on street engines.

As one cam grinder (Isky) put it: "A blower kind of covers up the holes in a motor." In other words, the blown motor isn't as touchy about several aspects of engine tuning, including cam timing and profile. In many cases a moderate street blower will work fine with the stock cam in the engine (as long as you install heavier valve springs). A stock cam would certainly work better than a typical high performance cam. The only exception, as we noted earlier, would be cams in late-model engines which have been retarded to keep the exhaust valves open later in order to promote more complete combustion of unspent hydrocarbons in the exhaust manifolds (for smog control). Such cams can be advanced in most engines by adding an indexed, multi-keyed crankshaft timing chain sprocket, or by drilling the dowel pin hole oversize in the camshaft timing chain sprocket and installing an offset timing pin bushing.

Although most aftermarket cam companies make several full competition blower grinds, hardly any list anything specifically for street supercharged engines. Most recommend their RV-type cams or their "mileage" cams for street blower applications, and this is a sound recommendation. For a very general rule of thumb, you would want something in the 260° to 300° (advertised duration) range, or not over 240° as measured at .050-inch lifter rise, with net valve lift in the neighborhood of .450-.500-inch. These are strictly ballpark figures, but most moderate street blowers would be happy with a cam in the 260-280 degree (advertised) duration range; bigger engines with bigger blowers could use a bit more.

One of the few street blower manufacturers who has done quite a bit of experimenting with camshaft timing and profiles for blown engines is Jerry Magnuson. Through his testing, he has developed a cam and kit which he sells to go with the MC-220 blower on small block Chevies. Its specifications are 224° intake, 226° exhaust duration (at .050-inch), .480-inch net lift, on 114° lobe centers. This cam was used in a 350 Chevy for a series of dyno tests for Hot Rod Magazine with the Penco, B&M, Paxton, Magna Charger, and Mach 2 blowers—all limited to six pounds of boost—and it produced excellent low-end power, an exceptionally flat torque curve, and good power through 6000 rpm.

The Magnuson cam is manufactured by Schneider Cams of San Diego, California, a well-respected cam grinder (especially in circle-track racing) and they explained the basic theory behind this, or any, good street blower cam. First, the cam should have minimal overlap, which means that the lobe centers should be spread as far apart as possible—in the neighborhood of 112°-115°. Overlap is the period during the cam timing cycle when the intake valve has already opened, and the exhaust valve is not yet closed. In a normally-aspirated engine such a situation is tolerable—in fact preferable—because it takes a certain amount of time to get the air moving. But in a blown engine the air is under pressure. As soon as the intake valve opens, the air/fuel mix bursts into the cylinder, filling it rapidly. If the exhaust valve remains open, part of the air/fuel mix will be pumped right out the exhaust pipe by the blower. This is obviously a waste of fuel and power.

Overlap can be decreased by decreasing the duration of both the intake and the exhaust valve cycles. However, even though a blown motor doesn't need as much cam timing as an unblown motor, you still want to keep as much duration as practical to fill the cylinder as much as possible, and to get the exhaust out. Widening (increasing the cam timing between) lobe centers reduces overlap without changing the length of time each valve remains open. However, doing so advances the exhaust valve timing and retards the intake valve timing; or, to look at it another way, it retards the intake valve timing relative to the exhaust valve timing.

Consequently, to get the intake valve timing back where it should be, Schneider recommends advancing the entire cam for a supercharged engine slightly (2 to 4 degrees). The net effect is to get both the intake and the exhaust valves open early—to get the pressurized air/fuel mix in quickly, and get the consequently pressurized exhaust also out as soon as possible—and at the same time to *close* each valve as soon as possible so that none of the cylinder pressure can leak back out. If the intake stayed open too long, the beginning compression stroke of the piston would want to push the "super-charge" back out the intake valve, similar to reversion in a normally-aspirated engine, except that it would have the effect of "backpressure" on the blower, which would ultimately reduce its mechanical, volumetric, *and* adiabatic efficiency. If the exhaust valve stays open too long, as we said, part of the new incoming air/fuel charge will blow out the exhaust port.

Finally, Schneider usually cuts street blower cams on a dual pattern profile, giving the exhaust more duration than the intake (as much as 10° to 12°). Widening the lobe centers and advancing cam timing also helps to make this possible without negative effects. The reason for giving the exhaust more timing on a supercharged engine should be obvious—the blower pumps a whole lot more air and fuel into the engine on the intake stroke, but the exhaust has to get out on its own. Many people (including us, come to think of it) have stated that the supercharger will actually help to "blow" the exhaust out of the cylinder, but this isn't wholly accurate. The blower will increase overall cylinder pressure, since it will pack a greater mass of air and fuel into it. After combustion, this "super charge" of air and fuel expands, filling the cylinder with a greater mass, and greater pressure, of hot exhaust gas than would be present in a normally-aspirated engine's cylinder (that's why you put the supercharger on the engine, remember?). Consequently, when the exhaust valve opens, the significantly increased pressure of the heated gases in the cylinder forces them out the exhaust port at a greater velocity. However, there is more mass of exhaust *in* the cylinder, so it will still take more time to get it all out. Make sense?

Now, just to throw another contradiction at you—if you remember back in the chapter on blower theory we were talking about mean effective pressure and the thermal efficiency of a supercharged engine versus a normally-aspirated one. Although we ultimately get much more power out of a supercharged engine (because its MEP is higher), nevertheless its thermal efficiency is lower than that of the unblown engine. The reason is that all of the extra energy pumped into the engine in the form of fuel by the supercharger simply cannot be converted to heat, then to pressure, and then to work within the cylinder...before the exhaust valve opens. The very fact that the exhaust rushes out of the blown engine under greater pressure, and at a higher temperature, than that in other engines tells us that a portion—a greater portion—of the potential power is being wasted. Of course all internal combustion engines have terrible thermal efficiency, supercharged or not, so this is not a major point. But to a certain extent the thermal efficiency, and a fraction of the power, of the blown motor is determined by the opening time of the exhaust valve. The sooner the exhaust valve opens (the more advanced the exhaust timing), the lower the thermal efficiency, and the more potential work (power) goes out the exhaust pipe. Like we said earlier, the supercharged engine is a melange of tradeoffs.

However, this is a minor point offered primarily as food for thought for those of you who like that sort of thing. The engine builders who *know* supercharged engines still agree that blower cams, in general, should have relatively short duration, wide lobe centers, and slightly advanced overall timing.

As far as the rest of the valvetrain is concerned, other than the importance of increasing valve spring pres-

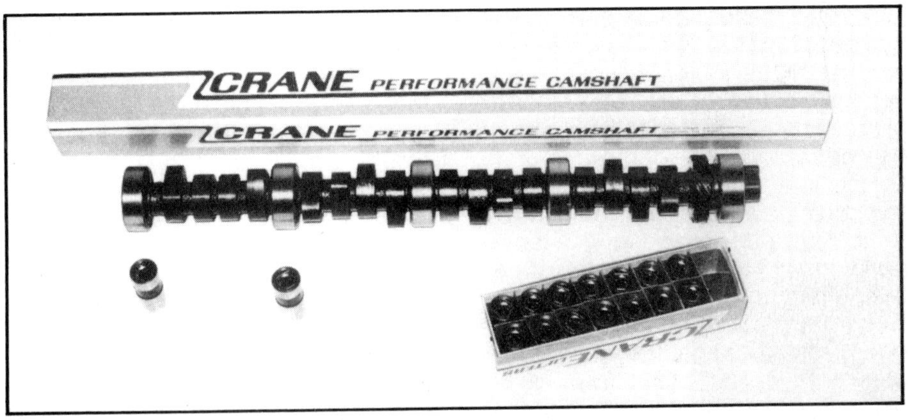

For a healthy street blower, such as a 4-71 or 6-71 GMC, you want a cam that won't strangle the inlet with short duration, but won't blow the boost out the exhaust with long duration and big overlap. Since few cam manufacturers offer specific grinds for such applications, a good compromise is a moderate street performance hydraulic cam such as the Crane Blazer.

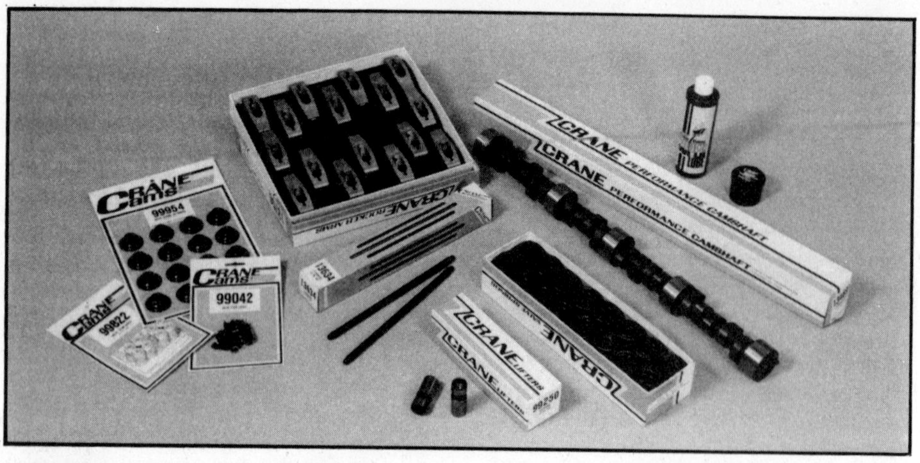

Once again, if you are planning to make some strong horsepower with your street blower, you will have to choose a camshaft and sturdy valvetrain components which are up to the task. For a healthy 6-71 blown street 454 Chevy, Art Chrisman chose a complete Crane valvetrain assembly including roller rockers, blue-stripe dual springs, solid lifters, and a 290-degree (CC-290-NC) "Commander" profile camshaft.

sures (as mentioned in the cylinder head section), you can use the same parts in a supercharged engine that you would use in any other performance engine. Hydraulic lifters should be fine for most street blown motors, but be sure to use good-quality anti-pump-up type (most are these days). Mike Kuhl prefers to use flat tappet cams in his high performance supercharged motors, to be *sure* that a lifter won't pump up, hang an intake valve open, and grenade the blower. But Mike's "River Rat" blown gasoline big blocks are far from what you would call the typical street supercharged engine.

CARBURETION

There has been lots of talk (magazine print) lately about "mapping" carburetors for blower use. In my discussions with dozens of blower experts, the only conclusions I can draw about specific carburetor preparation for supercharged engines is that there is little general concensus. So, at the outset of this section I would like to make one point very clear—trying to prescribe any sort of carburetor tuning in a book like this is exactly like asking a doctor to cure your ills over the phone. If your symptoms are common enough, the doctor might be able to phone-order you the proper prescription to make you better. But without physically examining you in person, the doctor can never be sure of exactly what's wrong with your particular body.

The same goes for carburetors—in fact for the entire "tunable" package of an engine, which includes the fuel and ignition systems, as well as cam timing and profile, valve adjustment, blower drive ratio, and several other factors. And of course the tuning of any particular engine—blown or otherwise—depends on factors such as vehicle weight, transmission type, rear axle ratio, rear tire diameter, and the type of driving you normally do...not to mention local elevation, typical weather conditions, and several other factors. Are you getting the picture? Unfortunately the hot rod market has come to expect exact answers from magazine articles and Tech Question columns, as well as "pre-tuned" bolt-on components that will work out-of-the-box on their performance engines. Such answers and components can be correct *up to a point.* But after that, it's up to a knowledgeable and qualified tuner to put his hands and test instruments on your individual automobile to get it properly dialed in. So what we say about carburetors here are only broad guidelines, or the sometimes subjective opinions of certain experts.

DRAW-THROUGH SYSTEMS

Superchargers can be set up in two basic kinds of induction systems. In a draw-through configuration, the blower mounts between the carburetor(s) and the intake ports, and it pulls air through the carb(s) from below, or from the outlet side. In a blow-through system, the carburetor (usually just one) mounts between the blower and the intake ports, with the blower pumping dry air into the top, or inlet, of the carb. The vast majority of today's street superchargers are of the Roots-type, and they are almost universally mounted directly to the engine's intake manifold, with the carburetion on top—a draw-through configuration. Blow-through designs currently pertain only to the centrifugal Paxton and Mach 2 superchargers.

Speaking very generally, the beauty of a carburetor is that it is a *responsive* device. The basic function of a carb is to *read*, or sense, the volume of air passing through it, and to deliver a proportional amount of fuel. The more air going through the carb—for whatever reason—the more fuel the carb delivers. That's how a venturi works. The problem is (1) today's typical 4-

The most common carburetor arrangement on street blowers is the draw-through set-up, with the carburetors sitting on top of the blower. This is a typical Supercharger Engineering installation, using dual Carter AFB's. Note fuel pressure gauge mounted at the fuel block (arrow), which will indicate the effectiveness of the fuel pump in keeping up with the blower's increased fuel demands.

barrel carburetors are filled with pumps, passages, nozzles, and wells that may not operate directly by the venturi effect, and are dialed in for normally-aspirated engines, and (2) some carbs, or some parts of some carbs, do not feed a great enough quantity of fuel to feed a healthy supercharged engine. Remember that the horsepower output of any engine is directly proportional to the quantity of fuel that it burns. If your supercharged V-8 is set up to make 600 horsepower, its fuel system—including the carburetors, fuel pump, fuel line, filters, and so on—must be capable of delivering at least twice as many pounds per hour of fuel as a 300-hp engine.

But don't let the last statement worry you. Most of our 4-barrels can deliver plenty of fuel. NASCAR engines pump out over 600 hp with a single carb. Dual 4-barrels on top of a blower should be able to handle the horsepower requirements of any street machine. However, the *proportion* of fuel to air delivered by the carb(s) on a blown motor may need to be adjusted.

Very simply stated—and universally agreed among blower experts—you must richen the carburetion on a supercharged engine. Why? First, in most carbs the blower seems to be able to pull air through faster than the carb is designed to deliver fuel. But the next two reasons are much more important. Second, in any high performance engine making a lot of power, you do not want to get into a lean situation since this will lead to extremely high cylinder temperatures, detonation, burned valves, backfiring, and other problems. Obviously you do not want the engine to run overly rich, which will foul the plugs and make the mixture harder to ignite. But given the choice between the two options—especially in a street engine—you would certainly prefer to be on the fat side rather than the lean side. The big problem in a blown motor is that, with so much air being pumped through, it can lean out severely in a big hurry, and you'd rather be safe than suffer a meltdown. And third, since all blowers (and especially Roots-types) heat the air charge as they pump it, it is practical and preferable to add extra fuel to the airstream to act as a coolant in the intake charge. Evaporating gasoline will significantly counteract the poor adiabatic efficiency of many blowers, and the over-rich supercharged engine will actually show more horsepower on the dyno. If you're worried about fuel economy, you shouldn't be bolting a blower on your motor.

The big difference between street blowers and competition blowers, from a visual standpoint, is that street blowers use carburetors while race blowers run injectors. To get the competition look on a 6-71, you can mount an injector-type scoop atop the carbs. Better yet, an in-line carburetor such as the Autolite shown here looks almost like a Hilborn 4-holer, but drives like a street machine should.

How much should you fatten up the carb on your supercharger? Most all blower manufacturers agree that 20-25% richer on the secondaries and about 10% richer on the primaries, plus a turn or two on the idle screws, will put you in the ballpark. Many of the engine builders say this is all you need to know about blower carbs. The main thing is to fatten it up to start with so that you don't hurt anything the first time you fire the motor. Then drive the car and listen for detonation, and pull the plugs to see how the mixture looks. If the engine sneezes when it starts, or has an off-idle stumble, you might be able to fix it with richer idle settings or a fatter pump shot. If the car runs strong but the plugs look sooty, you can try leaning it down a bit—but pay very close attention to the plugs for any signs of metal particles on the porcelain or electrodes, which would indicate detonation in the cylinders (in a blown motor it is sometimes difficult to hear detonation because of all the other noise...a careful and continual reading of the plugs is often a better guide). If you are familiar with carburetor tuning procedures, and you know the ins and outs of the particular carb(s) on your blower, you can basically dial in draw-through blower carbs the same way you would tune the carb for proper driving and power characteristics on any engine—you'll just end up with richer settings. If you aren't familiar with carburetors, take the car to someone who is, and let them tune it. However, *do not* expect to take a carburetor straight out of a box, bolt it onto your blower, and expect it to work properly.

How about some specific recommendations for a starting point? Supercharger Engineering offers the following suggestions for street carburetion on 4-71's, V6-71's, and 6-71's. They personally prefer Carter competition-type AFB 4-barrels (750 cfm, #4762 front, #4760 rear with mechanical choke). On these they suggest 10% larger primary jets, 20-25% larger secondary jets, and primary metering rods #16P241 with .071-inch and .047-inch steps. They also recommend changing float bowl needles and seats to larger .120-inch diameter, and stretching the metering rod springs to match the distance between the two idle-mixture screws on the carb, as a handy measurement. (Heavier springs are difficult to get; however if the engine still stumbles off-idle, install heavier metering rod springs if you can find them.) Finally, set the idle screws 2-1/2 to 3 turns open. If you prefer Holleys, Supercharger Engineering recommends #1850's (600 cfm) with the #65 primary jets enlarged to #70 and the secondary plate drilled to a #36 drill size.

Now, if you ask a *carburetor* expert—as opposed to an engine builder or a blower manufacturer—how to set up carbs for superchargers, you're bound to get a lot more answers...many of which may apply to carburetors in general rather than specifically to blower carbs. For this purpose we spent some time with engineer Mike Jones, who operates a highly-respected performance carburetion shop in Huntington Beach, California.

Mike agrees that the first necessity is to richen a carb for blower use, primarily to guard against detonation. But he also points out that a big blower, like a GMC, acts almost like a huge plenum below the carb—like a tunnel ram. The mixing action of the rotors effectively atomizes the fuel, so you don't have dispersion problems with most blowers (that is, you get even fuel distribution to all the cylinders). But the blower also has the negative effect of leaning out the idle mixture, so that the engine desires more fuel just off idle. Consequently, most blower carbs require richening of the off-idle circuits. Also, Mike feels that most carbs have poor intermediate circuits. They hold too little fuel in the idle circuit well to begin with, so when a supercharger greatly increases the airflow through the carb, this sucks the intermediate circuit fuel supply dry. Consequently you feel a "flat spot" as you accelerate.

Other particular problems Mike ascribes to individual carbs by brand. Which is the best carburetor for blower use? In terms of airflow capacity, the Holley looks best. Mike states that, at one atmosphere of boost (approximately 14.7 pounds), a Holley will just about double its cfm airflow (that is, there will be no airflow loss through the carb). An AFB under the same conditions will cause about 20% airflow loss, and a Quadrajet will reduce airflow close to 40% because of frictional losses through the venturis. However, Mike is quick to point out that a typical street engine (350 cubic inch) simply cannot use much more than 900 to 950 cubic feet per minute of intake volume, so even with a single 600-750 cfm 4-barrel on top of a 6-71, you're not likely to reach the point where the carb can't flow as much air as the blower can feed to the engine.

So, in choosing between Holleys and AFB's, Mike says that the AFB is better for blowers "because the main circuit comes in so quickly." In the AFB the booster sits higher above the venturi than it does in the Holley, so it will pick up the vacuum signal from the venturi more quickly. Mike, who is not one to mince words, states that Holleys "are usually fixed with band-aids, because they are designed with crutches." What he means is that performance tuners often cover up a problem area in a Holley by modifying some other area. Another dislike Mike has for Holleys is the number of rubber components in the carb, which are easily attacked by today's highly aromatic gasoline blends (of course, this has nothing to do with superchargers). However the AFB has only one rubber part, the accelerator pump, which should be replaced with a Buna-N pump to avoid deterioration. Mike admits that many tuners apply "band-aids" to the AFB as well, the primary one (and this does apply to blower applications) being the use of "huge" secondary jets. Mike says the proper adjustment is to add weight to the secondary air door bob weights, rather than simply enlarging the jets.

A close-up of another Autolite in-line carburetor installation on a 6-71 street blower shows how neatly it fits the blower—and a Hilborn scoop. Unfortunately these carbs are now quite rare, having been offered during the '60's for Shelby Fords.

Since the Holley and the Carter AFB 4-barrels are by far the most popular choices in carburetion for today's blowers, we have included a complete photo-essay on each one, showing typical modifications as recommended by Mike Jones.

Surprisingly enough, however, Mike personally likes the Rochester Quadrajet 4-barrel and states, "I firmly believe the Q-jet would work excellently on blowers for the street." The Quadrajet's primary advantage, according to Mike, is that it is "infinitely adjustable." Its one drawback is that it has a relatively small float bowl, so you need a high-volume fuel pump to make sure that it doesn't starve for fuel. Even better than the Q-Jet, says Mike, would be the Thermo-Quad, which has most of the same qualities as the Q-jet, but also has a large fuel reservoir and an air-door secondary. Says Mike, "The 800 cfm Thermo-Quad would be excellent with a blower...a killer."

Injector-type aluminum scoops to fit 4-barrel carburetors are made by several of the blower manufacturers. Blower Drive Service offers these two to fit dual or single quad carbs.

HOW TO REBUILD HOLLEY CARBS FOR STREET BLOWERS

As we said in the beginning of this section, there is no such thing as a mail-order tune-up. We can't begin to give you specific recommendations on how to set up the carburetor(s) on your particular blower on your particular car. Any number of variables will affect carb tuning with street blowers, including the type of blower you have, the amount of boost it makes, other components in your engine (especially the camshaft and ignition timing), the weight of your car, the type of transmission it has, and so on. However, there are some specific areas in Holley 4-barrels that could use some attention when mounted on a blower, and there are some basic procedures you can follow while attempting to get your fuel system dialed in. We visited Mike Jones' carburetor shop in Huntington Beach, California, to get the following pictures and recommendations for adapting Holleys and AFB's to blowers (in a draw-through application). His approach may differ from others, and the suggestions given here are simply rough guidelines.

Mike selected this 735 cfm 4150 double pumper Holley as a good choice for a GMC blower, especially if one carb is used. It has a secondary metering block with removable jets, and vacuum-operated secondaries, which Mike prefers for blowers so "the engine can decide when the secondaries should cut in." The following steps will apply to most Holleys, though.

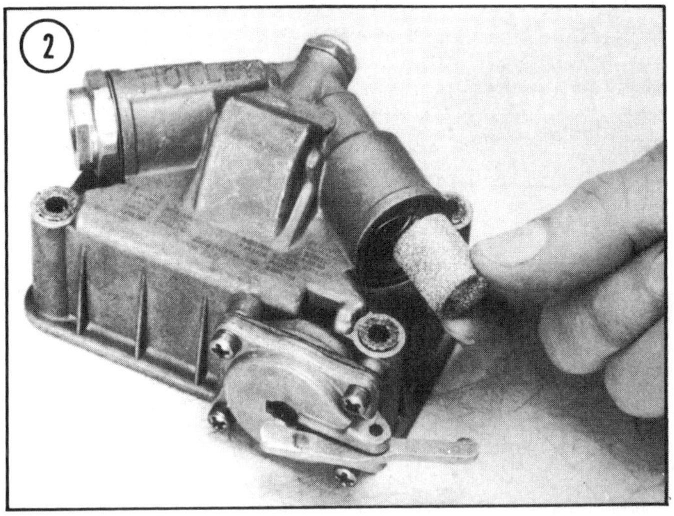

The first step with any Holley is to remove the bronze fuel filter in the float bowl—it is *very* restrictive. Replace it with a good, high capacity in-line filter.

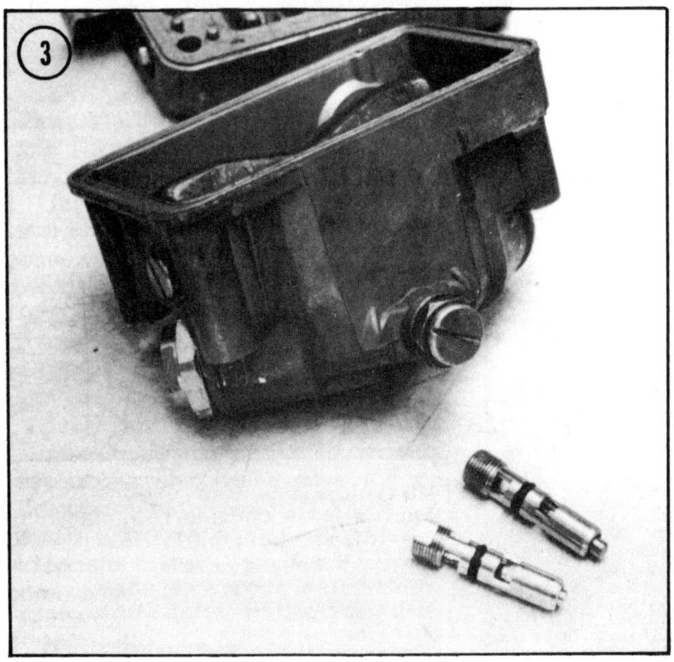

A blown motor is going to need more fuel than a stocker. Install a larger needle and seat, such as the 6-515 (bottom) with .130-inch orifice or 6-502 with .120-inch orifice (top). However, you don't want to overcome the float with fuel pressure, so start with the 502 first. If there is evidence of fuel starvation, then go to the 515.

As explained in the text, the suction of a supercharger tends to "dry up" certain circuits of the carburetor, often resulting in a bog, or "flat spot," either just off-idle, when the secondaries cut in, or whenever you stomp on the accelerator. One of the first things that many tuners do to cure this problem on Holleys is to install a larger accelerator pump kit, as shown. In many cases this will work—and it is much simpler than the alternative steps which follow—but Mike calls it a "Band-aid" rather than a cure. If you do use this fix, Mike recommends the 50cc pump, no. 20-11, and says to use the brown cam (the yellow cam is for Dominators). Also be *sure* that the larger pump housing does not bottom on the carb adapter plate—a common problem which bends or breaks the carb base.

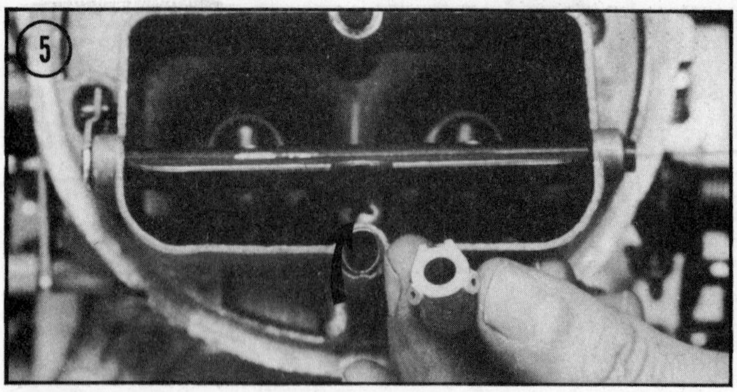

Rather than installing bigger pumps first, however, Mike says to start with the pump nozzle (or "shooter") located under the screw denoted by the arrow. Supercharged cars should never use pump nozzles with tube-type squirters, since the blower will constantly pull fuel out of them (that is, they will dribble). Most blower Holleys can use the "straight type" of nozzle, shown on the left in the close-up (denoted by the prefix 121-). If you are using a small carb on a big engine, or with lots of boost, you may need the special "anti pull-over" nozzle designed specifically to stop dribbling in high venturi signal situations (denoted by prefix 125-). Aside from making sure the blower doesn't siphon fuel out of the pump well, the idea here is to increase the orifice size of the nozzle before enlarging the pump. If the carb stumbles, try a .005-inch larger shooter. As a ballpark figure, Mike recommends #121-135 (.035-inch) in a 735 cfm carb; #121-131 (.031-inch) in 600's.

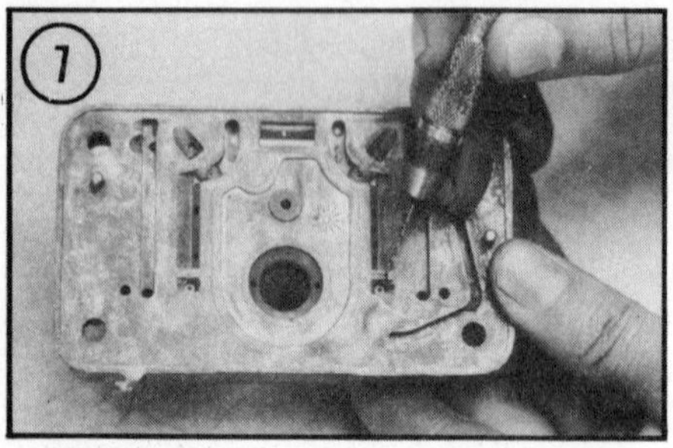

Now, if your car still has an off-idle stumble, here is the proper (though more tedious) way to correct it. Remove the primary metering block, and drill out the idle feed restrictor orifices, as shown, .002-inch at a time until the stumble clears up. If the metering block in your carb has emulsifier tubes instead of restrictor holes, you can't enlarge them (but you could switch metering blocks).

To go with the larger orifices, you need a larger power valve, as shown on the left compared to a standard Holley valve (center) and a typical rebuild kit valve (right) which have obviously smaller outlets. The "High Flow" Holley valve is denoted by prefix 125- followed by a 1 and two digits which indicate the vacuum, in inches of mercury, at which the valve opens. That is, a 125-185 valve opens at 8.5 inches of vacuum. To determine which valve you want, take a vacuum reading *at the base of the carb* (not in the manifold, where your vacuum/boost gauge is hooked up) with the car idling normally. Then select a valve which opens at least 3.5 inches below the vacuum reading; i.e., if the reading is 10 inches, use power valve #125-165, which opens at 6.5 inches of mercury.

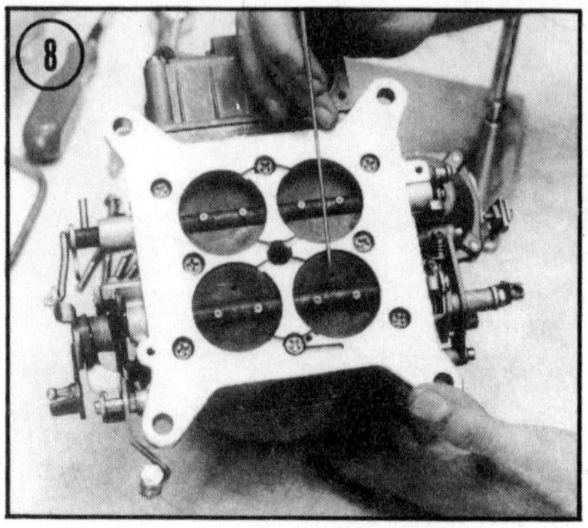

Another reason why Holleys have an off-idle stumble on high performance or supercharged cars is because the small shot of fuel in the curb idle slots dries up when the butterflies are kept slightly open to maintain a higher idle speed. In order to keep the throttle blades covering the idle slots, while still allowing a healthy idle, drill a .093- to .120-inch hole in each primary throttle blade as shown to allow air to pass through at idle. Start with .093-inch, and increase if necessary.

If by chance you drilled the idle restrictors too large (which is uncommon), or if your car tends to "load up" at idle (which is rare with a blower), you can correct this by enlarging the idle air bleed holes, as shown, .003-inch at a time (to a maximum of .009-inch larger).

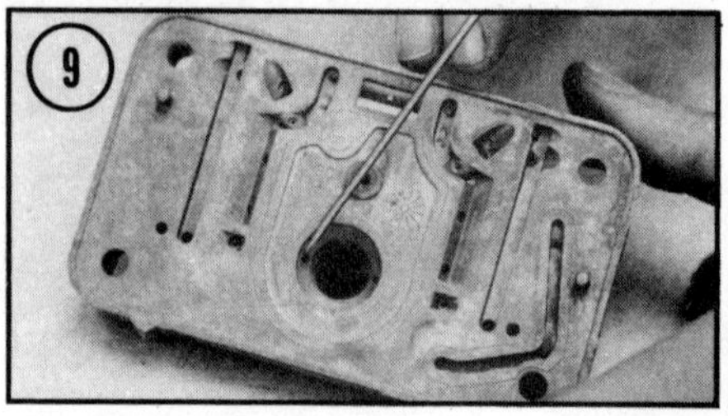

Most tuners tell you to increase main jet sizes in blower carbs right off. Mike prefers, in Holleys, to start by increasing the power valve flow instead, since this will give you more fuel when you need it, and save gas while you're cruising (plus it will also help cure the acceleration bog). Start by enlarging the power valve restriction holes, as shown, .005-inch at a time (Mike typically enlarges them about .015-inch total). However, don't drill the holes larger than .098-inch each, since available power valves won't flow any more fuel than this anyway.

After you have made the above changes, *then* it is time to consider changing main jets. Mike's approach to Holleys is to open up the power valve restrictors, play with different power valve settings, and then *moderately* re-jet the carb to get the best combination of power, smoothness, and economy. Using a Holley chart to convert jet numbers to thousandths of an inch (drill size), Mike now recommends increasing primary main jets about .005-inch, and secondaries about .008-inch. "If you have to go much over this," says Mike, "Let's pay attention to the power valve restrictors instead."

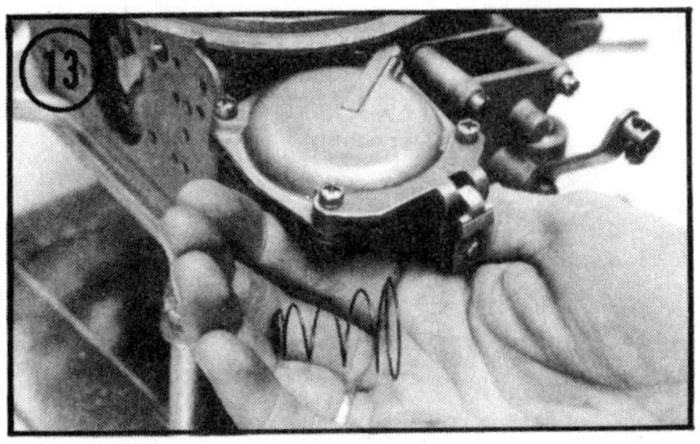

Before you begin similar operations on the secondary metering block, first get the vacuum diaphragm dialed in with the proper tension spring. A common cause of secondary bog in vacuum-operated Holleys is that the diaphragm spring is too weak, allowing the secondaries to open too quickly. To "tune" the secondaries, get diaphragm spring kit #20-13 (which has seven color-coded springs) and install successively weaker springs until the bog disappears. However, Mike says that "80% of the time the purple spring is the right one."

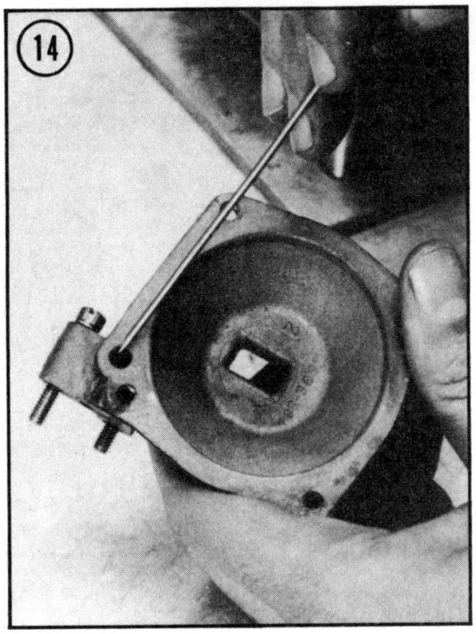

If the diaphragm housing on your carb has a 3/16-inch check ball in the orifice shown, discard it. It is a secondary dampener, which is not needed in blower applications.

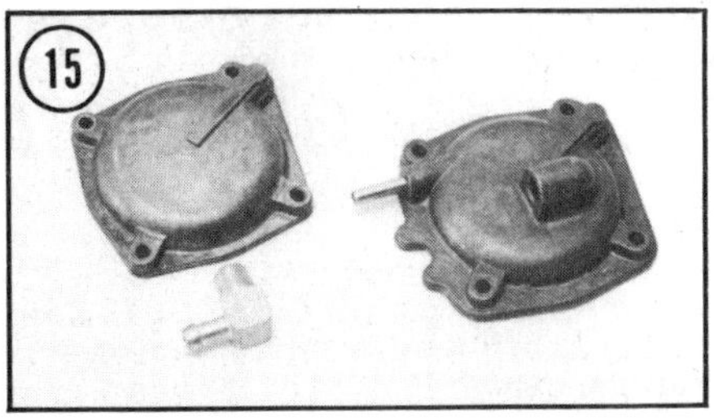

If you run dual vacuum secondary Holleys on your blower, you *must* run a vacuum balance hose between the two diaphragm housings. A Holley diaphragm top with a built-in tube for an equalizer hose is shown at right, but these are getting rare. Mike simply drills and taps the top of a standard housing, and inserts a 1/4-inch fitting for the hose.

If your Holley has a secondary metering plate, like this, rather than a block with removable jets and a power valve, Mike suggests getting a secondary block kit from Holley (#34-13). This is the best route, but they are expensive. If you prefer to keep the secondary plate, however, make the following modifications to increase fuel flow. After the proper secondary spring has been installed in the secondary diaphragm, open up the idle feed restrictors, as shown, .005-inch at a time, to clear up any remaining secondary bog. (If your carb has a secondary metering block, follow the procedure shown in step #7 instead.)

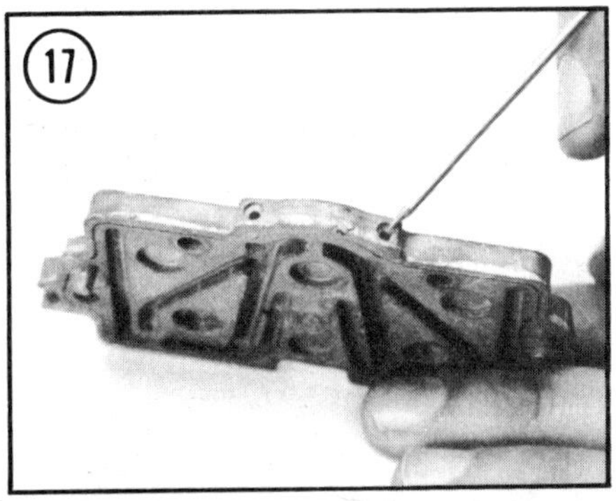

These holes in the secondary plate are the same as secondary jets. Mike suggests drilling them about .010-inch larger for blower applications—you need to go larger on them than you would with removable jets since the plate is so restrictive. Obviously a metering block with removable jets is preferable for any performance application with a Holley.

Whether you have a secondary block or plate, if you enlarged the secondary idle feed restrictors too much, or if you are getting a stumble accompanied by a puff of black exhaust smoke when the secondaries cut in (that is, an overly rich rather than a lean situation), you can drill out the secondary idle air bleed holes, as shown, in .003-inch increments.

Obviously the few steps shown here can't make a Holley expert out of you, nor are they guaranteed to solve the problems of your carb(s) on your blower. (S-A Design does sell an entire book, like this, just on Holley carbs.) Plus there are plenty of other things you can do with Holleys to make them better for blower applications. For instance, Mike built the carb shown as a special high performance blower unit. It uses a 700 cfm main body on an 850 cfm base to get smaller venturis (to pick up fuel signal quicker) with larger butterflies, for a total of 760 cfm. This carb also has bar-type boosters placed high in the venturis, plus 4-corner idle circuits. Mike states that "This is our best blower carb," and they can be ordered, complete, through his shop.

HOW TO REBUILD CARTER AFB'S FOR STREET BLOWERS

Mike Jones' approach to AFB's is similar in theory to that for Holleys, but the design of the carb itself differs considerably. All AFB's are basically the same. In blower applications, the biggest problem with standard AFB's is that the largest secondary jets available are too small and the smallest primary metering rods available are too large. These problems should not affect dual carb set-ups, but for single AFB blower applications, Mike says the solution is to use a Competition Series AFB, either #CS-4760 750 cfm, or #CS-4758 and #CS-4761 500 cfm (for dual carbs).

Our example carburetor is an electric-choke 500 cfm AFB, but all are basically the same.

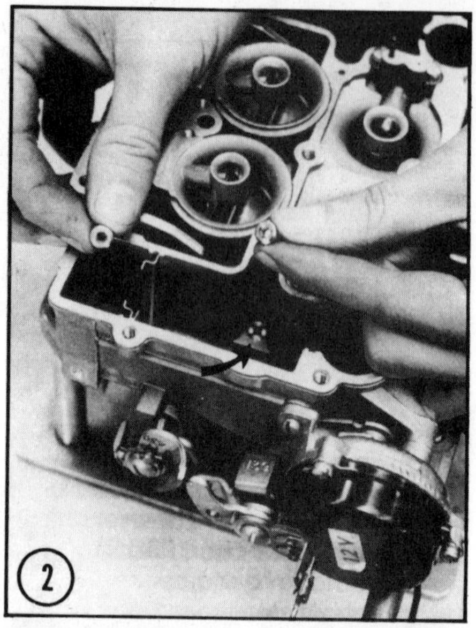

Normally main jets do not need changing to tune an AFB, but in a blower application they do. Mike suggests starting with .003- to .005-inch larger primary jets. Once you have the primaries dialed in, increase the secondaries the same amount. The jets are located at the bottom of the float wells (arrow).

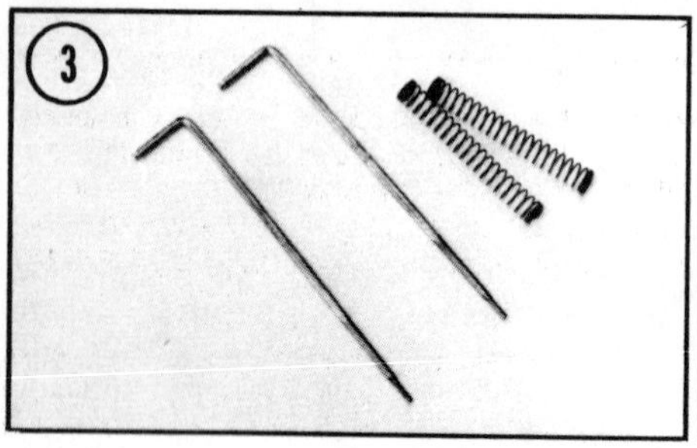

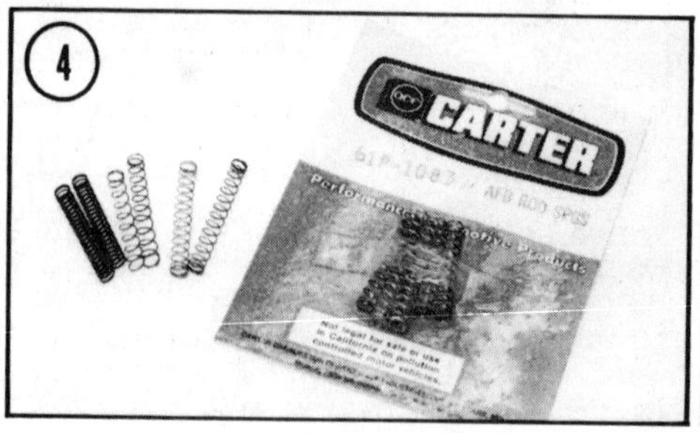

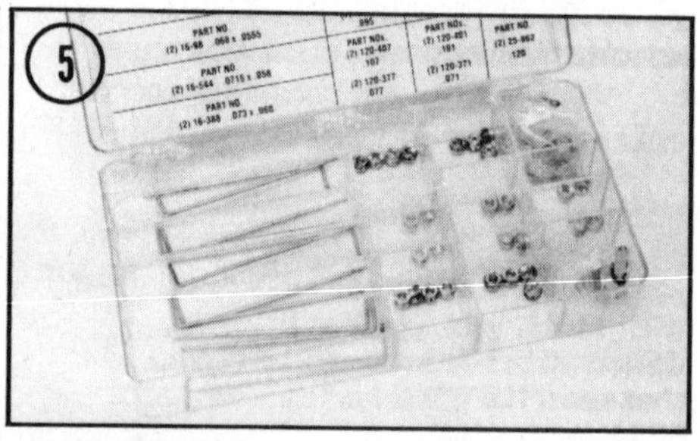

In an AFB fuel is metered through the primary main jets by the metering rods, which have a stepped outside diameter. As the rod moves up in the jet, it allows more fuel to pass. Springs balance the movement of the metering rods against carburetor vacuum, so that they deliver fuel on demand. The tip diameter of the rod determines maximum primary side fuel delivery at wide open throttle. The stiffness of the spring determines how soon the rod will fully lift. To tune the carb, you should get a Carter Strip Kit, which contains several sizes of metering rods and main jets, plus you need a metering rod spring kit (#61P-1083) Start with the smallest-tip metering rods (#16-241) which are the richest, then lean the carb back in increments until it bogs, and go one step richer. On the springs do likewise. Start with the stiffest spring (richest); install weaker springs until it bogs; then go back to the next stiffer springs. This will give you the best balance of performance and economy.

In a blower application, an AFB might still have an off-idle stumble, or a bog on acceleration. If so, remove the accelerator pump nozzle, as shown, and drill the tips of the tubes larger in .002-inch increments (to a maximum of .006-inch larger).

If this doesn't solve the problem, remove the primary venturi clusters as shown...

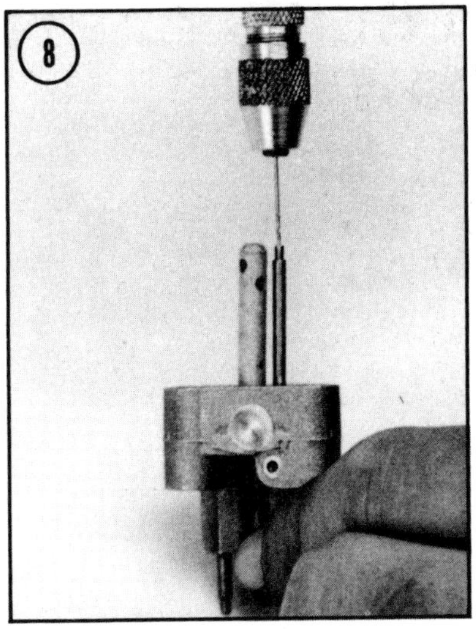

And drill the idle fuel jet tubes larger in .002-inch increments until the stumble clears up.

To produce a smooth transition into the secondary circuit, without a lean spot, the AFB uses a pair of auxiliary air doors (arrow) located in the secondary venturis just below the venturi clusters. A pair of weights on the ends of the air door shaft keep these butterflies closed until suction through the secondary throttle bores overcomes the leverage of the weights. The problem on performance applications is that a stronger secondary signal pulls the air doors open too quickly, which causes a flat spot, or bog, when the secondaries open. Obviously this problem is greatly intensified when a supercharger is pulling air through the carb.

The problem is that heavier-weighted air doors are not available from Carter. However, if you dig through some older AFB's in the junk yard, you will find some with larger, heavier weights. The heaviest, shown on the left, came on '58-'62 Chryslers. If your AFB's have a secondary stumble, look for heavier air doors.

To solve the air door problem in AFB's, Mike is having special weights cast in lead, aluminum, and magnesium to allow easier secondary tuning. These throttle door kits are now available from Mike Jones Carburetion in Huntington Beach, CA.

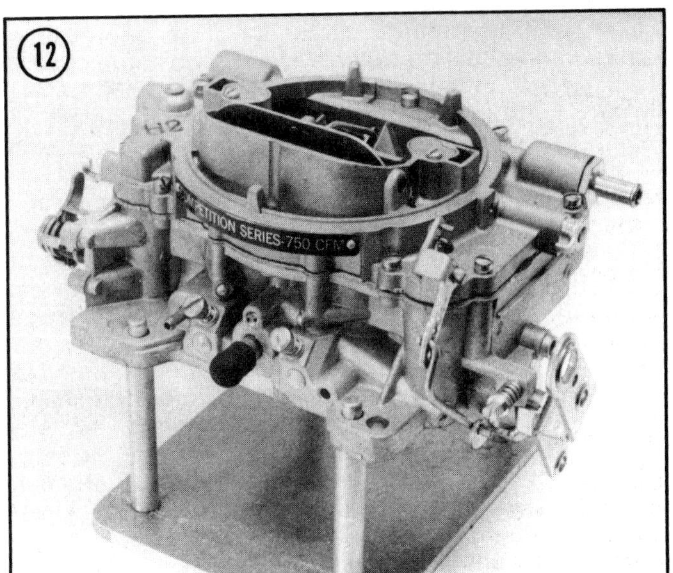

If you find that your standard AFB still runs too lean on your blower even with the richest jets and metering rods available, the best solution is to switch to a Competition Series AFB, such as the 750 cfm model shown here.

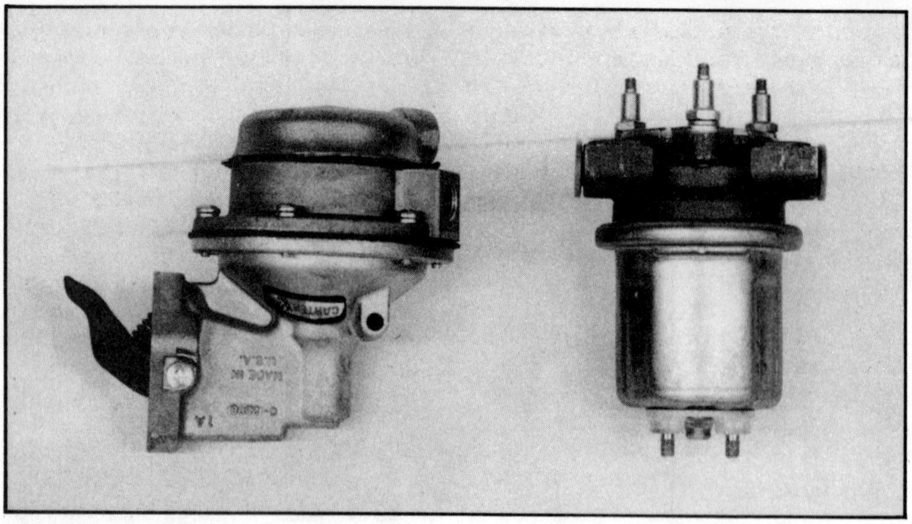

If a supercharger increases the power output of an engine, you will have to increase the fuel input to the engine. For draw-through systems, fuel volume is more important than fuel pressure, and a good high-volume mechanical pump, such as the Carter shown on the left, should be more than adequate. In blow-through systems, however, fuel pump *pressure* must always exceed blower pressure by at least 3 psi, or else fuel will not flow into the carb. In such cases, an electric pump, such as the one shown on the right, will probably be necessary.

As we briefly mentioned, the rest of the fuel-supply system on a supercharged vehicle is very important, also. There is little point in reworking the carburetor(s) to supply more fuel if the fuel pump, fuel lines, or filters cannot deliver that quantity of fuel to the carb. Under no circumstances do you want a supercharged engine to lean out from fuel starvation, especially under full throttle at high rpm. Such a situation can result in ventilated pistons in a big hurry.

Each of the components in the fuel-delivery system is equally important, since a restriction at any point between the gas tank and the carburetor float bowl will affect the entire system. In other words, a 1/4-inch fuel line from the tank isn't going to cut it on a 600 hp engine. However, even if you replace the entire line with 1/2-inch tubing, but leave the same pickup, screen, and outlet fitting in the tank, you have gained nothing.

The fuel pump is, of course, a critical component in the fuel system. Electric fuel pumps are popular since they are easy to install (and they leave extra room at the front of the engine, where the mechanical pump previously mounted...which often helps when you are trying to mount accessories, along with a blower drive, on the engine), they are often adjustable, and they can be mounted near the fuel tank to insure constant pressure throughout the system. There are a couple of drawbacks to electric pumps, however (not counting the fact that they stop if the wire breaks, if they short out, or if they encounter some other electrical malfunction). First, they deliver fuel at the same rate no matter how fast the engine is running. Obviously the engine needs less fuel at slower speeds. Consequently the electric pump puts greater demand on float bowl needles and seats to seal (and therefore such pumps should never be used without a good, adjustable fuel regulator at the carburetor—preferably one for each carb for multiple setups). Second, if the needles do not seal, of if some other leak occurs in the system, or if your vehicle should for some reason ever find itself upside down or in some similar predicament, the electric pump continues to deliver fuel at an alarming rate until the electrical power source is cut off. These are certainly not common occurrences, but they can happen (I say this from experience).

The point is, if you have room for a mechanical pump, you might as well use one. They are more reliable, their delivery rate increases proportionally with engine speed, and they're usually less expensive. Mike Jones recommends them highly, saying "I haven't seen a mechanical pump that couldn't outflow an electric pump at high speed." He says good choices are the Delco fuel injection Corvette pump for small block Chevies, or the Carter competition street pump (not the high pressure version).

Mike also strongly suggests removing built-in fuel filters from any carburetors, since most are terribly restrictive, and adding a good, high-flow filter to the fuel line. Good canister types are available from Fram and Holley, among others, and excellent aircraft in-line filters are available from most suppliers of braided stainless hoses. By the way, Mike also strongly discourages the use of rubber, including hoses, at any point in the fuel system—especially if you might be using a toluene-based octane boost additive—since the aromatics in today's gasolines attack rubber, cause it to break down, and then clog fuel passages.

How can you judge if your fuel supply system is adequate or not ...without waiting for pistons or valves to burn if it isn't up to snuff? The easiest and surest way to monitor fuel supply is to install a fuel pressure gauge as close to the carburetor fuel inlet as possible. If you have a GMC on top of your motor, and the carbs stick through the hood, you can mount the gauges right at the carbs so you can read them from inside the car. Otherwise, you could mount fuel gauges on the cowl, or in some other external location. It is not wise to put a fuel pressure gauge inside the passenger compartment, however, since a pressure line filled with gasoline must be connected to the gauge—and you wouldn't want something like that leaking inside the car. You only need about 3 pounds of fuel pressure at each carburetor. As long as the fuel system maintains this pressure at all rpm levels, and under heavy loads, you have nothing to worry about. If the gauge indicates a pressure drop at high speed, however, especially under load, then you need to increase the supply or search for restrictions.

BLOW-THROUGH SYSTEMS

If you could leave the carburetor, intake manifold, distributor, and so on, in place on your engine, add a supercharger on a front-mounted bracket like an alternator, and then run a big air pressure hose into the top of the carburetor from the blower—that would be a pretty simple way to supercharge your motor. Such a setup is known as a "blow-through" system, whereby the supercharger pumps dry air *into* the carburetor from the top (or into the inlet side). Unfortunately, most of our currently popular superchargers (Roots-types) do not lend themselves well to blow-through installations. So, other than a few obsolete types of blowers, or home-made installations, the only modern superchargers which could be mounted in a blow-through configuration are the Paxton and the Mach 2. Of these two, the Mach 2 is most often sold as a draw-through setup, while the Paxton in most cases *must* be installed in a blow-through configuration. (Certain new Paxton blowers, fitted with special shaft seals, can be used in draw-through applications...no other Paxtons will tolerate fuel inside the blower.) Consequently, most any particular information you may need about blow-through carburetion systems can be supplied to you by Paxton or Mach 2, along with your blower.

A blow-through design certainly simplifies, and speeds up, the blower installation process. And, to be quite honest, most applications that we have seen, at moderate boost levels, seem to work quite well without any fussy tuning. However, blowing compressed air into the mouth of the carburetor does pose some problems.

On the theoretical side, the compressed air passing through the carburetor venturis when the blower is making boost will be denser than the same quantity of air being sucked through a top-mounted blower carb. That is, the mass of air is greater, but the volume is smaller. The problem is that carburetor venturis sense air velocity, which is dependent on the volume passing through the venturi. Consequently, the carburetor would not feed the proper proportion of fuel into this denser air stream. Of course, the carb jetting can (and should) be increased to keep the air/fuel ratio compatible with blower boost. But on a street-driven car, the blower only gets into boost a small percentage of the time. So if the carb is jetted for dense air, it will be too rich under low-boost conditions. Now remember we are talking theoretically here. In actual practice, a blow-through carb can be set richer on the secondary jets (assuming you are using a 2-stage carb, such as a 4-barrel), which will come into play only when you are getting full boost.

A second theoretical problem with blow-through systems is that the dense air blowing through the carburetor will not be as effective at atomizing the fuel being injected into the airstream by the carburetor, and this problem is compounded by the fact that the carburetor sits very close to the intake manifold (in the stock location), into which the air/fuel mix is forced at pressures above those for which the manifold was designed. In contrast, a Roots-type blower drawing air through a top-mounted carb produces better fuel vaporization, and then further mixes the fuel and air before pumping it almost directly into the ports. Oftentimes a blow-through blower design is chosen specifically to keep the blower and carb under a low hood line. However, such low-profile blow-through systems can sometimes upset manifold design characteristics and lead to poor fuel distribution. On a recent test with a Paxton blower, a Holley carb with a bonnet, and an aftermarket dual-plane intake manifold we saw severe distribution problems (one plane of the manifold being way too rich, the other plane being way too lean). A carb spacer, a better dual-plane manifold design, or perhaps an open-plenum or X-type manifold would have worked better. As for manifold selection on a particular engine, we suggest you consult the blower manufacturer. And it wouldn't hurt to read your plugs a few times after some hard-boost passes to make sure you aren't running too lean on some cylinders.

Another peculiarity with blow-through systems is that the carburetor and fuel system must be "pressurized" to match blower boost. If you simply piped pressurized air from the blower into the mouth of the carburetor, what would happen? In the first place, if the carb fuel reservoir (float bowl) were vented to atmosphere, the main jets would deliver no fuel into the air stream, since the venturi effect relies on a pressure *drop* at the venturi to siphon fuel from the bowl. If there is pressure in the carb throat, instead of vacuum, it would blow fuel out of the carb, instead of drawing it into the airstream. However, most of our popular carburetors have float bowl vent tubes which open into the top of the carburetor mouth, so that they would be pressurized along with the air going through the carb. If such is the case (it is with Holleys, AFB's, and most carbs), a proportionate pressure differential will be maintained between the bowl and the venturi, so the carb should work properly.

If there are any other vents, holes, or possible leakage areas in the carburetor body, these must be sealed. A bonnet-type blow-through set-up will also test all carburetor gaskets: mating surfaces must be flat, screws must be tight, and gaskets should be in good condition. One area that will always leak on bonnet blow-through carbs are the throttle shaft bores. On relatively low-boost applications, with a new carb which doesn't have worn shafts, you might not have to worry about the amount of air that leaks here (although it can be messy, since it will blow fuel out, as well as air). The only sure cure is to cut a groove in each throttle shaft bore and install a small O-ring in each before re-installing the

The major advantages of a blow-through supercharger system are that the intake manifold and carburetor can be left in place, and that the blower can be mounted in a variety of places in the engine compartment. However, blow-through systems—whether they pressurize the carburetor with a bonnet, as shown here, or in a complete box—do require slightly more complex tuning and modification of the fuel system.

shafts. This obviously isn't an easy task. An easier quick-fix used by Paxton is to drill a small hole into each shaft bore from the bottom of the carb base plate, and then add a thin spacer plate beneath the carb which is drilled to match these holes, with intersecting horizontal holes drilled into the center opening of the plate. This will route manifold pressure to each throttle shaft bore, to equalize the pressure coming from the carburetor throat.

Another—and usually better—way to pressurize the carburetor in a blow-through application is to place the entire carb inside a sealed pressure box. These are available from Paxton to fit several types of carbs and engines. The box is made in two pieces, with the bottom part mounting between the carb and manifold. Fuel line, throttle linkage, and any other carburetor connections must be routed into the box through air-tight seals. Then the top is attached, and the pressure hose from the blower is attached to it. Not only does a "blower box" equalize the pressure on the carburetor, inside and out (thus requiring no sealing of the carb), but it also places a plenum of pressurized air around the carb from which it can feed. With bonnet-type applications, especially if the bonnet must sit low over the carb mouth, the column of air may not flow evenly through the carburetor.

With either a bonnet or an air box, a final requisite for blow-through systems is an increase in fuel pump pressure. Since the float bowl must be pressurized to match blower boost in a blow-through carb, consequently fuel pump pressure must *exceed* blower boost at all times, by approximately 3 psi. If it doesn't, the pressure inside the float bowl will be equal or greater than the fuel pressure, and fuel would not be able to enter the bowl. Consequently, if the blower will make a maximum of nine pounds of boost, you will need at least a 12 psi fuel pump. To get this sort of pressure, you will have to use a good electric pump, preferably an adjustable type. It would also be wise to install a good adjustable fuel regulator, and a fuel pressure gauge so that you can adjust it properly. Finally, if the carburetor has hollow floats, especially brass ones, they will probably collapse under pressure. Replace them with closed-cell synthetic foam (i.e. Nitrophyl) floats, which are readily available for most carbs.

Of course, all carburetors for blow-through applications must be re-jetted richer, as any blower carb does. In general, Paxton says "make sure the car is carbureted properly to begin with, before the blower is added. Then enlarge the secondary jets approximately 20 per cent." Of course, this is strictly a starting point. Paxton recommends Carter AFB 4-barrels for most of their applications, suggesting the 500 cfm #9500 for engines up to 305 cubic inches, and the 625 cfm #9625 for engines up to 350 inches. To these carbs they suggest installing .098-inch primary jets and .104-inch secondary jets, and setting the float level at 9/32-inch. For bonnet-type installations, they also recommend quickening the accelerator pump shot. Paxton's small block Ford blower box is designed for a "flat-top" Autolite 4-barrel, but will also accept a vacuum secondary 600 cfm Holley. On this carb they suggest installing the stiffest diaphragm spring available for the secondaries to prevent a bog. On some blow-through applications, particularly bonnet-type, it is difficult to get vacuum-operated secondaries to work properly.

FUEL INJECTION

The only arguable reason for running fuel injection on a street supercharged engine is for the looks or for the effect. If that's what you want, there are systems being marketed today that make it workable. But there's nothing a fuel injector is going to give you on your street blower that a couple of carburetors won't—other than mid-range and off-idle flat spots and general lack of drivability.

If you really want a big bug catcher or a four-port Hilborn on top of your 6-71, first consider the expense—not just of the injector itself, but also of the necessary pump, pump drive, valves, hoses, and tank. Then consider where you are going to mount said pump, drive, hoses, and so on, and still retain street necessities such as a water pump, alternator, radiator, etc. Finally, consider that these types of competition injectors are designed to be operated in two driving ranges only: idle and *go.* If you're still interested, such systems can be adapted to street use, and actual street-injection blower systems are offered by Blower Drive Service.

The BDS "Street King" injection systems are available in a variety of sizes, from 2-holers for V-6 blowers to 4-ports and bug catchers for 6-71's. The key ingredient for making the system "streetable" is a fuel reservoir tank, fitted with one or more Holley float chambers, which can be mounted above the injector pump to keep it primed. This eliminates the need to get out and squirt fuel into the injector with a priming can before starting the engine.

However, if you would prefer the practicality and the drivability of a carburetor with the looks of a 4-port Hilborn atop your 6-71 Jimmy, here's

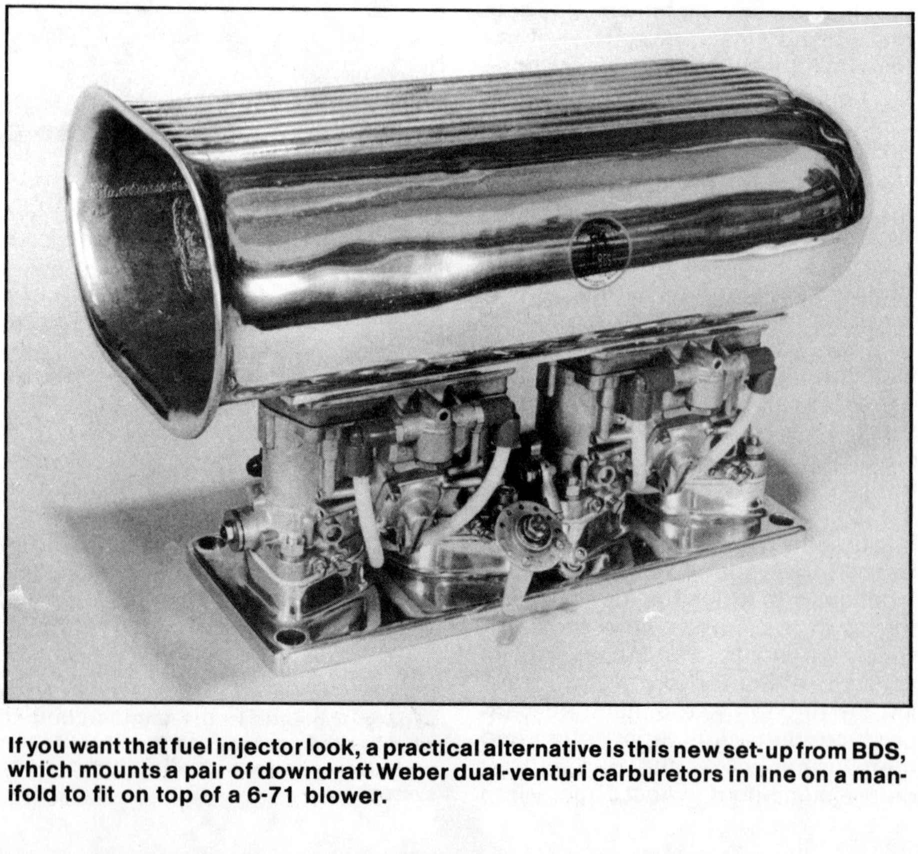

If you want that fuel injector look, a practical alternative is this new set-up from BDS, which mounts a pair of downdraft Weber dual-venturi carburetors in line on a manifold to fit on top of a 6-71 blower.

Inglese Induction Systems of Branford, CT, makes adapters to mount side-draft 40mm to 55mm Webers on blowers. The adapters actually bolt onto 4-bbl carburetor plates, so they can be readily attached to most any blower top. Inglese's systems come complete with linkage, braided hoses, air cleaners, and any other necessary parts.

one alternative. Try to locate one of the "inline" Autolite 4-barrels made in the late 60's and sold through the Shelby parts program. This carb is similar to a pair of 2-throat Webers hooked together, and it looks very much like an injector with a Hilborn scoop on top. They are quite rare, but we have seen at least a couple on street blowers.

Of course, the other alternative is to mount a pair of downdraft Webers, Mikunis, Dellortos, or similar carburetors in line on top of the blower. These carbs are also expensive, but at least they are readily available. And BDS has recently introduced a cast aluminum manifold which will mount a pair of these carbs, in line, on a 6-71.

Since we are discussing carburetor exotica, we should not close this section without mentioning other types of carbs. Webers, Dellortos, Mikunis, and the like are excellent carburetors, they are perfectly applicable to blowers since they are nearly infinitely adjustable (both in terms of jetting, as well as in terms of venturi size, or cfm capability). However, a full discussion of these types of carburetors is unfortunately beyond the scope of this book. Magna Charger MC-220 blowers come with a pair of side-draft Dellortos. When we called the U.S. Dellorto distributor for further information on these carbs, they told us, "Ask Jerry Magnuson; he knows more about them than we do." Magnuson has done extensive testing and research with these carburetors on superchargers—both on automobiles and on motorcycles—and would be the best source of information on them. For Weber carburetion for superchargers, we would recommend you contact Inglese Induction Systems of Branford, Connecticut.

IGNITION SYSTEMS

The ignition system in any vehicle has only two functions: to ignite the charge, and to ignite it at a prescribed time in the engine's cycle.

Of these two functions, obviously the primary one is make the spark plug fire. Any time that a spark plug fails to fire, or the engine "misses," no power whatsoever is made in that cylinder on that stroke. In an 8-cylinder engine, one misfiring plug will reduce horsepower more than 12 per cent. What would cause a spark plug not to fire? Resistance—either in the distributor cap, in the plug or coil wires, in the plug itself, or between the plug tip and electrode (the plug "gap"). Of course, a total break anywhere in the secondary electrical circuit—such as a disconnected plug wire—is complete resistance.

Voltage is a measure of an electrical current's capacity to overcome resistance. An automotive ignition system (the secondary circuit) converts battery voltage (12-volts) into approximately 40,000 volts in the coil, and then "distributes" it to each of the spark plugs in the firing sequence. The reason for the high voltage is to make the current jump the gap from the spark plug tip to the electrode, which makes a big spark. The distance between the electrode and the tip—the plug gap—determines the amount of electrical resistance between these two terminals. A given amount of voltage will bridge a given plug gap in thin air.

The problem is that the cylinder isn't filled with *thin* air. In the first place, the air has fuel in it; in the second place, the air is hot; and in the third place, at the exact time the plug needs to fire, the air/fuel charge has been considerably compressed by the piston. In other words, the plug needs to fire in very *thick* air, and this causes a considerable increase in resistance at the plug gap. You could reduce the size of the gap to reduce the resistance, but this would reduce the size (or "fatness") of the spark produced, which would reduce its effectiveness at lighting a fire in the fuel/air charge.

Consequently, "high energy" performance ignition systems have been developed which increase the voltage in the secondary circuit. This allows you to run the same, or even a wider, plug gap to produce a bigger, fatter spark with more assurance that the spark will overcome the resistance at the gap, and fire, on each power stroke. Some of these systems use a higher-voltage coil, some generate their own high voltage (magnetos), and the most modern rely on electronics to produce a very rapid, fat spark. Some, such as the MSD, even produce several rapid sparks—to make sure at least one lights off the mixture—in the short duration of a previously conventional spark.

It is not our purpose here to describe, and to argue the merits of, the several types of high performance ignition systems available on the market these days. You can find plenty of literature on these elsewhere (for instance in S-A Design's PERFORMANCE WITH ECONOMY or COMPLETE GUIDE TO BOLT-ON PERFORMANCE books, see page 128 in this book). The only consideration in this respect is that *a supercharger increases the density of the air/fuel charge in the cylinder.* Not only does the blower make the air in the cylinder even "thicker" than that in a normally-aspirated engine, but most blown en-

Voltage is what makes electricity jump a gap, and a blown engine under boost needs plenty of ignition voltage to get the plugs to fire in the dense atmosphere of the combustion chamber. The easiest way to increase ignition system primary voltage is to install an aftermarket high voltage coil.

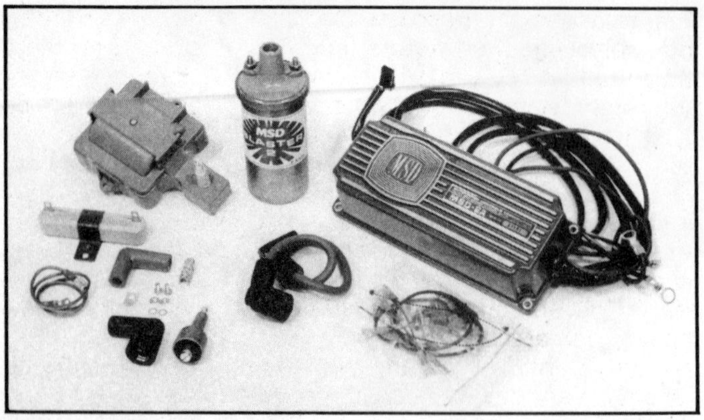

One of the better high performance electronic street ignition systems is the MSD-6A, which hooks up to a standard distributor. The MSD not only increases spark voltage, but fires several sparks in short duration to make sure one lights the fire in the cylinder.

gines will also be tuned to run somewhat richer, as well, which means that the resistance will be further increased, the "wetter" mixture will be even harder to ignite, and the plug tips will have more tendency to foul.

So, in theory, a supercharged engine should require a high performance ignition system to (1) make sure the spark can overcome the increased resistance at the plug gap, and fire on every power stroke, and (2) produce a fatter spark to get the denser mixture ignited. The type of ignition you choose is not all that important; in many cases, especially with GMC blowers, a primary consideration will be the physical size of the distributor. Rear-mounted distributors, such as on small block Chevies, need to be relatively small in diameter to fit behind a 6-71 (in other words, a GM HEI distributor won't fit). Engines with front-mounted distributors, such as most Fords and Buick V-6's, cause greater clearance problems with blowers. In many cases a horizontal-pole ("crab" type) distributor cap will give enough room; in severe cases, an angle-drive for the distributor or magneto is the only solution (just remember, when you are shopping for a blower system for your engine, be sure to figure out how and where the distributor is going to fit *before* you purchase and install all the other parts).

At what point do you need to upgrade your ignition system? It all depends on how much blower boost you are running, how rich your mixture is, how wide you have gapped your plugs, and how effective the existing ignition in your engine is. To give you one very revealing example, for a street blower test for Hot Rod Magazine a basic point-type distributor (a Mallory YL dual-point unit) and a *stock factory Chevrolet coil* was used for dyno runs on several blowers up to 6000 rpm at up to nine pounds of boost. At no point in any of these tests was any ignition misfiring encountered, as confirmed on an oscilloscope. Frankly, this is surprising, but the fact is that regular old points-and-coil ignitions work pretty well. There appears to be no need for a hotter ignition at the 6 to 9 pound boost level. Naturally, ignition requirements increase with higher boost levels.

The true test for any ignition system, however, is to run it and see if it works. If you do encounter misfiring, begin by checking the various components in the system: the rotor, cap, plug wires, terminals, and the plugs themselves. A visit to a good diagnostic shop, with a test scope, will pinpoint ignition system trouble spots quickly for you. If everything is in good shape, and it still misfires, then you can figure you need to upgrade the system to produce a bigger spark. Either the ignition lights the fire or it doesn't—that's all there is to that part of the ignition discussion.

The second part of the discussion revolves around *when* the distributor delivers the spark to the cylinder. This gets a little more complicated, since it's not a yes or no question. However, the basic rule on ignition timing for performance engines is that you want to advance the timing as much as possible, and you want any variable ignition timing (advance curve) to "come in" as quickly as possible, *without causing detonation.* This one condition—causing detonation—is what makes ignition timing on a blown engine different from that on other performance engines.

There are three variables in ignition timing. Initial advance is determined by where you screw down the distributor in the engine. Turning the distributor body one way advances ignition timing relative to camshaft and crankshaft timing, turning it the other way retards it. Centrifugal advance is built into the distributor. As distributor speed increases, a pair of sprung weights in the distributor body pivot outwards by centrifugal force, and advance the rotor position in the housing. The amount of centrifugal advance is determined by the position of stops on the weights; the rate of centrifugal advance is determined by the size of the weights and/or the stiffness of the springs they work against. Finally, most stock distributors also have a vacuum advance mechanism built into the distributor as well. This system retards ignition timing, usually by rotating the breaker plate, under low carburetor vacuum conditions, such as during acceleration. Initial advance is your basic "overall" ignition timing adjustment. Centrifugal advance fires the plugs sooner at higher rpm's to make up for lag time of the burning process at higher engine speeds, plus allows the engine to get started. And vacuum advance (or more accurately "retard") reduces pinging or detonation under acceleration or heavy load. (NOTE: The distributor turns at half crankshaft speed. Centrifugal and vacuum advance are usually given in "distributor degrees," while initial advance is given in crankshaft degrees. One distributor degree equals two crankshaft degrees. Thus an ignition system which has 10° of centrifugal advance in the distributor and 10° of initial advance at the crank, has 30° of total ignition timing advance, as read in crank degrees.)

Opinions vary on ignition timing recommendations for blown engines.

Racers, especially those used to building fuel or alcohol motors, such as Mike Kuhl, do not want to run any ignition retard (since retarding the timing hurts performance). Kuhl says he prefers to "lock out" any advance mechanism in the distributor or magneto; or, if any curve (centrifugal advance) is retained, to make it come in as quickly as possible. Most street blower builders, however, recommend retaining the centrifugal advance, slowing down the curve a bit, and setting the initial advance very conservatively. For instance, Paxton recommends for their blower kits, using 91-octane gasoline, to retain 9° of centrifugal advance, have it "all in" by 3400 rpm, and to set initial advance at 8° BTDC, for a total of 26° of timing. Magna Charger recommends limiting centrifugal advance to only 8°, having it all in by 1600 rpm, and setting initial lead at 16° on leaded fuel, 14° for unleaded, with a 32° total timing overall limit. They also recommend retaining the stock vacuum advance. Supercharger Engineering suggests 10° of distributor advance, in by 3000 rpm, with 18° initial lead for a total of 38° for street, less (10°-14°) initial lead for racing. Isky cams' T. Willie Hutchins states that a blown motor doesn't need more than 25° of total ignition advance, and that "modern ignitions have too much lead."

As you can see, there is a certain lack of consensus among the experts. This is understandable, since each engine acts a little differently, the total engine package will be different in most every case, and people drive their cars differently—all of which alter ignition timing requirements.

A major consideration is that engine tuning—whether on blown motors or others—is a package deal. You can't set the ignition timing without taking into consideration the carburetor tuning, or the cam timing, and so on. Carburetor expert Mike Jones made this point emphatically when we asked him how to set up carbs for street blown engines. The first thing he answered was, "Let's see the distributor." To clarify, he explained that 80% of blower tuning problems revolve around detonation—you tune the engine for maximum performance up to the point of detonation, but then you must back off. Since the onset of detonation is directly influenced by ignition advance and curve (as one element), other tuning procedures—such as carburetor settings—must be symbiotic with the ignition.

Mike recommends setting up the distributor first, and offers several concrete guidelines. For street blown motors, on today's gasoline, he likes to slow down the distributor curve considerably, usually by installing stiffer springs on the centrifugal weights. His recommendations are to start the centrifugal advance curve at 1000-1200 rpm and have it all in by 3500-3800 rpm. On a small block Chevy he likes 12° advance in the distributor, with 8°-14° initial. Ford and Mopar small blocks like about 4° less initial advance, while big block Chevies like 12°-14° in the distributor with 4°-8° initial lead. In general, Mike says an across-the-board recommendation for any street blown engine would be between 10°-12° in the distributor and 8°-14° initial advance, which gives you a range of anywhere between 28° and 38° of total ignition advance.

Mike further recommends retaining a vacuum advance system on all street blown engines, with the vacuum line plugged into the carb, not into the manifold. However, the increased air velocity through draw-through carbs may give too much vacuum, which could be compensated for with an adjustable vacuum canister, such as are now available from Accel and Crane Cams. On a blow-through system, vacuum advance would not work. Finally, Mike reminds us that the centrifugal advance curve in the distributor should not begin below engine idle speed. In other words, if your engine idles at 1200 rpm, don't have the advance curve coming in at 800 rpm. Doing so, especially on cars with automatic transmissions, will often cause the engine to stumble or die when the car is put in gear.

If you do install a high performance ignition system on your engine, especially one with greatly increased voltage to the plugs, it would be wise to install better-insulated, higher-capacity wires and boots to make sure that the electricity doesn't find some path of lower resistance than the plug gap. Also, to take full advantage of most high-energy ignitions, you should open the plug gap somewhat to allow for a fatter spark—however, follow the ignition manufacturer's recommendation in this regard. In a blown engine, you might not be able to use quite as much gap as a normally-aspirated engine can. Also, since cylinder temperatures are usually higher in a supercharged engine, you might need to go one step colder in spark plug heat range, especially for higher-performance applications.

These figures and recommendations are certainly no more than basic guidelines—beginning points. As we said earlier, there is no way that we can tune your engine for you in a book. It is strictly a trial-and-error process that every engine builder and car owner must go through (or have done) to make the entire package work to its optimum. If you do not have the equipment to perform such tuning operations at home, take your car to a qualified, performance-oriented shop that does (or take your distributor there before you install it).

HEADERS AND EXHAUST

Supercharged engines do not like backpressure. It's very simple. If you are pumping a greater quantity of air into the engine on the intake side, you've got a greater quantity that has to get out the exhaust side, and you don't want the plumbing backing up. If you plan to run your blown engine in any sort of competition where open exhaust is allowed, you definitely want an "openable" exhaust system. A supercharged engine picks up significant power when it is uncorked.

For the street, however, the object is to keep the exhaust system as unrestrictive as possible. A set of headers would be a good choice, especially on higher-boost blown motors. In the case of a supercharged engine, though, typical header "tuning" theory does not apply. In other words, the length, diameter, and collector style of the headers is of little consequence. The rule of thumb for blown motors is to use big tubing, since exhaust gas temperature is higher in a blown engine and it will expand more, and to keep the primaries short if you plan to run the headers open. You have undoubtedly noticed that supercharged drag cars never run headers that have collectors. They just use large diameter, short "zoomies" to get the spent gas out as quickly as possible.

For the rest of the exhaust system on your blown streeter, use as large-diameter tubing as practical (2 or preferably 2-1/2 inch diameter), keep the number of bends and the overall length to a minimum, and use low-restriction "turbo" type mufflers. The turbo muffler was originally designed for the turbocharged Corvair, specifically to reduce backpressure to a minimum on this production "blown" street machine. However, these days turbo-type mufflers are made by several companies, and they are not all the same. If possible, use mufflers which have 2-1/2 inch diameter orifices all the way through. Remember that an exhaust system is just like any other fluid-flow system—the smallest restriction at any one point affects flow through the whole system.

There's really nothing else to say about getting the exhaust out of a blown engine. Just remember that the capacity of the exhaust system—or the amount of exhaust the engine will have to expel—will be proportional to the amount of air the blower is pumping in...except that the exhaust is much hotter, and therefore takes up considerably more volume. For further information on backpressure, muffler flow capacity, and exhaust system modification we would highly recommend S-A Design's PERFORMANCE WITH ECONOMY by David Vizard, see page 128.

Today's street blower drives are more sophisticated than this one, but even the best leave little room for cooling system components. Electric fan(s), such as these, may solve one problem. Also note the two small hoses which carry radiator water from the top tank directly into holes tapped into the heads—necessary because the older blower manifold had no water outlets.

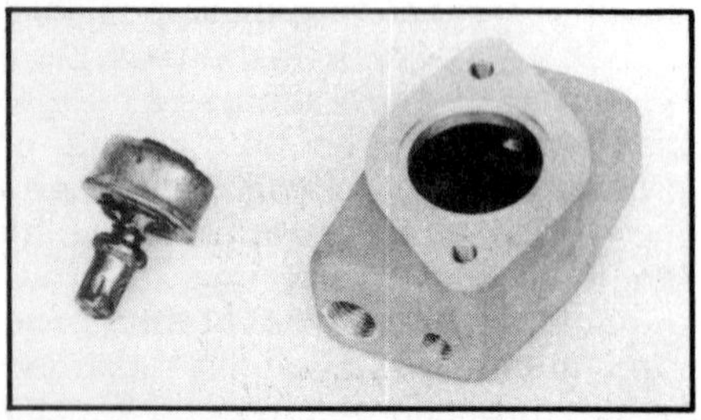

If your blower manifold has no water neck built in, and you must fabricate an upper coolant supply system, you can route two small hoses from the heads (or the front of the manifold) into this remote thermostat housing available from Offenhauser and most blower shops. It accepts a regular water neck, to attach a single hose to the radiator, plus it has provisions for attaching a temperature gauge sender and a heater hose.

THE COOLING SYSTEM

One essential component of the total vehicle power package often overlooked by performance engine builders is the cooling system. Any high performance engine will require increased cooling capacity, since roughly 30% of the engine's power is absorbed by the cooling system. Remember that the power (energy) made in the motor is first converted to heat before it is converted to work. The more power your engine makes, the more heat it produces, and the more heat must be dispelled by the radiator. A supercharged engine not only makes more power, but it also inherently creates more heat because (1) the blower heats the intake charge a certain amount and (2) a certain portion of the engine's power is used up in driving the blower.

Consequently, you are going to need a good, big radiator for your supercharged engine, and a good water circulation system to feed it. Just how much radiator you need obviously depends on several factors, including how much extra horsepower you are making in the motor, how well the radiator is ducted to fresh air (and how well hot air is ducted out of the engine compartment), and even ambient climatic conditions. We really can't tell you how big a radiator you need. All we can do is forewarn you that a supercharged engine is going to run hotter. If your stock cooling system is good, a 5-7 pound boost blower probably won't require any cooling system change (other than perhaps a lower temperature thermostat, especially in late-model smog cars). However, if your cooling system is marginal to begin with, the blower is only going to make things worse.

The cooling system situation is compounded in most supercharged vehicles because the blower drive takes up most of the room at the front of the engine occupied by the water pump, fan, fan shroud, and radiator. In no case should you consider eliminating the fan on a blown motor. If you can't fit a conventional fan between the engine and the radiator, mount an

If your blower set-up doesn't leave room for a standard water pump at the front of the motor, a solution might be a remote water pump, such as this one made by Don Hampton. It can be hung anywhere on the front of the engine, like an alternator.

aftermarket electric fan on the front side. If you don't have enough room for a good-sized radiator, do not try to get by with a thinner one just to make room for the blower drive. You might have to re-mount the radiator in the chassis, or have a custom radiator built to handle the situation. But don't skimp here. You plan to *enjoy* the supercharger on your engine, right? Well an engine that overheats is no fun at all.

Contemporary street blower kits make provisions for water pumps in various ways. Most are designed to allow retention of the stock pump in the stock location, driving the pump either with the blower belt or with an accessory V-belt. Most current street blower kits also incorporate a standard water neck and thermostat in the intake manifold. Earlier competition-type blower drives made no provisions for water pumps, nor did most have any water outlets in the manifold. If you want to use such a blower system on the street, or if you are fabricating a custom blower setup, you might find a remote-mounted water pump (such as made by Don Hampton or by Mike Kuhl) to be the perfect solution. If your blower manifold does not have a water outlet to return coolant to the radiator, you will have to either drill and tap approximately 1-inch holes in the fronts of the heads, or into the front of the manifold (if possible), insert pipe fittings, and run hoses to the radiator. If you want to use a thermostat (which is advisable for street), you can route the water lines to a remote thermostat housing (available from most blower builders, or from Offenhauser), and then to the radiator. Particular cooling system design will vary on custom applications; we illustrate a few examples here. However, if you must tap into the heads of your engine, remem-

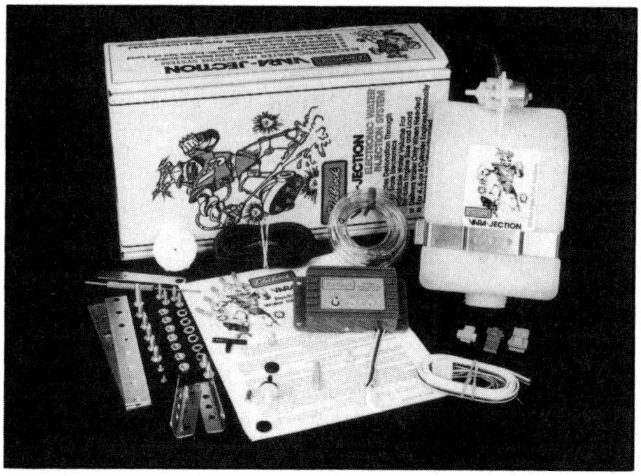

The Edelbrock Vara-Jection water injection kit is one of a few which will operate on vacuum or pressure, so it can be used with a supercharger. Its electronic module also senses engine speed to determine amount and timing of water to be injected into carb throat.

Another trick little water pump for blowers is this very small (about 4 inches wide) one made by Mike Kuhl. It mounts on a competition-style timing cover and is driven off the camshaft, like the fuel pump is on a race car.

ber that the path of water flow through the cylinder heads is critical to proper, and uniform, engine cooling. Insert water outlets only at the front of the heads (rather than anywhere in the middle), and do not make the openings too large if you do not use a thermostat in the system. Iskenderian offers a water-outlet drilling template for small block Chevies to accompany their 6-71 blower kit (which does not have a manifold water outlet).

BLOWER ACCESSORIES

WATER INJECTION

Although nearly all turbocharger kit manufacturers offer water injection units either as an integral part of the kit, or as an accessory, surprisingly few of the current blower manufacturers sell water injectors. Perhaps the main reason is that few of today's blower builders expect their customers to run unleaded, low octane gasoline in their supercharged engines.

In a blower comparison dyno test conducted for Hot Rod magazine, it was surprising to find that none of the blowers tested required water injection, or any other anti-detonant, when producing 6 pounds of boost in an 8.8 to 1 compression small block Chevy, using 92-octane leaded premium pump gasoline. In fact, some of these blowers made up to 8 or 9 pounds of boost at upper rpm ranges without showing signs of detonation. However, one blower which produced a significant amount of heat in the intake manifold did cause the engine to detonate severely, and even water injection did not solve the problem.

Whether or not your blown engine might need the addition of a water injector will depend on a variety of factors. Basically, they are the factors which influence whether or not the engine will tend to detonate, or "rattle," under boost and/or load. They include the static compression ratio of the engine, the amount of boost the blower makes, the amount of ignition timing advance, the heat produced in the cylinder by the blower (or other causes), carbon build-up on the pistons or combustion chambers, hot spots in the cylinders, and so on. The other major factor determining the point at which an engine will detonate is the octane rating of the fuel being used, but this is a more-or-less variable factor, rather than one built into the engine.

As we said earlier, detonation is the primary limitation to the amount you can supercharge a street engine. If you plan to drive your blown car a lot, and you want to be able to fill it up with pump gasoline (as opposed to buying racing gasoline, aviation gasoline, or "octane booster" additives), you may find detonation or pinging to be a problem even with relatively low boost levels. If the supercharger causes the engine to detonate, you might as well not have one, since an engine that is detonating loses torque and horsepower immediately. In other words, an engine that is detonating isn't making power. Water injection can inhibit detonation effectively, up to a certain point. But so can other means such as richening the fuel/air mixture, retarding the ignition timing, lowering the engine's compression ratio, lowering the blower boost (by reducing the blower drive ratio), or increasing the adiabatic efficiency of the blower itself (or devising some means of intercooling the charge).

What does water injection do? It simply cools the intake charge, primarily through evaporation of the water. It adds no power to the engine in itself. But it allows the engine to make power beyond the point at which it would

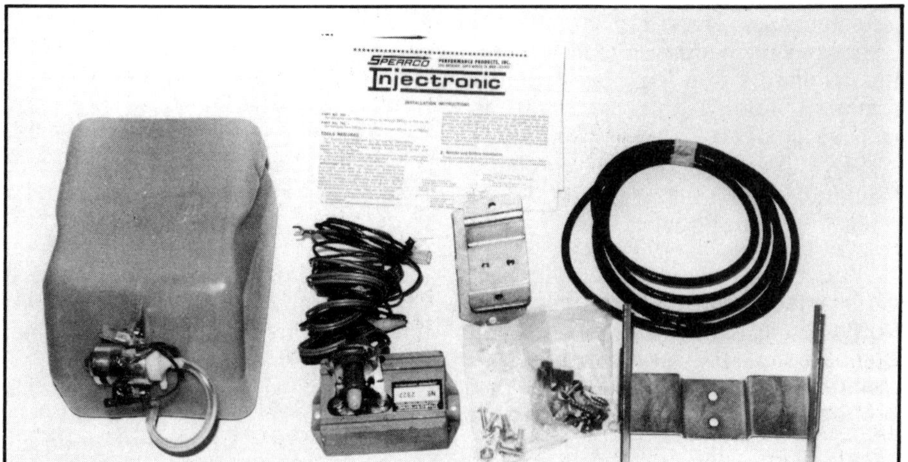

Several water injection systems have been developed for use with turbochargers, and these are perfectly applicable to superchargers, as well. An excellent unit of this type is the Spearco Injectronic system, which senses both manifold pressure and engine speed to determine water flow.

begin to detonate otherwise. Therefore water injection *allows the engine* to make more power. However—and this is an important point to remember—an engine will not make *as much* power with water injection as it would if the water injection were not needed under the same conditions. On the other hand, other methods to control detonation such as retarding the ignition, lowering the compression ratio, or limiting the blower boost each reduce horsepower as well. The only ways to really "fix" a detonation problem in a blown engine are to make the combustion chamber more efficient, make the blower more efficient, or raise the octane of the fuel.

So, figuring that you have a given engine, and you have a certain blower, and pump gasoline has only so much octane, water injection looks like a practical solution if your combination of parts does tend to detonate. Several types of water injectors are available on the market these days, and most any of them can be hooked up to a blown motor. The standard set-up includes a water container which mounts somewhere in the engine compartment, a small electric pump motor at the container to pump the water to the carburetor, a length of small diameter hose with one or two "jets" at the end to direct the water into the carburetor, above the venturis, and a pressure-actuated switch to turn on the pump motor when manifold pressure reaches a certain level. The only difference between a water injection unit for a normally-aspirated engine and one for a blown engine is the setting for the triggering switch: those for an unblown engine work at a low *vacuum* reading, those for a blown engine work at a set, or adjustable, manifold *pressure* level. Obviously, this pressure switch must be plumbed into the manifold, below the blower. Boost-actuated water injector switches, as well as complete blower-compatible water injection units, are available from most of the turbocharger manufacturers and retailers, if not from all blower outlets.

You could even make your own water injector from a window-washer bottle, pump, and nozzles; and you could use either of the two pressure switches used to light the yellow and orange boost dash panel lights from a Buick Turbo Regal or other model to actuate the unit under boost. However, if you are going to rely on a water injection unit to safeguard your blown motor from harm under boost, we would suggest you purchase a good unit that will be sure to deliver water when actuated, and will not clog.

A water injector should also be "tuned" to a given motor. First of all, the boost-actuated switch should be adjustable so that you can set it to operate at the point where detonation begins. You don't want to be dumping water into your intake charge when it isn't needed. Secondly, the water injector should be adjustable for rate of flow. You want just enough water in the airstream to stop the knocking, but no more (remember, the water is actually a contaminant in the air/fuel mixture). Gale Banks Engineering, and a few other companies, offer a water injector which comes with a series of color-coded restrictor jets (each having a different orifice diameter) which can be inserted in the water-supply hose to meter the amount of water that flows into the carb.

Finally, if a water injector is to be used on a blow-through supercharger system, first of all the pump motor must be able to pump at a higher psi than the boost in the airbox above the carburetor, the components in the system must be able to withstand pressure, and a check valve must be installed in the water line between the nozzles and the pump to keep blower pressure from bleeding back into the water cannister when the injector is not operating. Such a system is available from Paxton Products, as well as from several of the turbo manufacturers.

NITROUS OXIDE INJECTION

My first impulse is to say that a supercharged engine doesn't need nitrous oxide injection. A well set up supercharger will give your street machine all the extra horsepower it can use. Superchargers and nitrous injectors produce the same results in an engine, so you don't really need both.

On the other hand, blowers and nitrous injectors arrive at the same results by very different means. The blower physically compresses the intake air charge in order to make it denser, and heats it somewhat in the process. The nitrous injector makes the air charge denser by *cooling* it. Consequently there may be some very direct benefits from coupling a supercharger with nitrous.

Nitrous oxide injection was first used extensively during World War II on supercharged, piston-engined airplanes as an anti-detonant. If you really want to make some horsepower in your blown street engine, and detonation is an apparent problem, nitrous oxide injection might be a better (and certainly more powerful) solution than water injection. As we saw earlier, the greatest problem in supercharging is the creation of heat in the intake charge, which not only induces detonation and leads to parts damage, but also diminishes the blower's power return. If your street blower is set to make 6 to 8 pounds of boost, you shouldn't encounter any substantial heat problems, and you won't need anti-detonants.

However, if you are planning to pump 10 to 15 pounds of boost into your motor, especially on pump gasoline, you're going to have to do something about detonation and heat. Water injection or gasoline additives might take care of the problem. But nitrous oxide injection might be twice as good.

If you've ever driven a blown car on the street, and paid attention to the boost gauge, you have undoubtedly been impressed at how *seldom* the engine actually operates under a heavy boost load. The blower only "comes on" when you open the throttle all the way—and on most supercharged cars you just *can't* spend a whole lot of time at wide open throttle...it would be far too dangerous on city streets. Consequently, such an engine would need an anti-detonant only for short, relatively isolated, periods of time. This is the type of situation under which a nitrous system works best. If the nitrous were set up to actuate at full throttle only (as most are), it would not only cool the intake charge, thus proportionally lowering the blower outlet temperature, but would also inject extra fuel to richen the mixture under full boost—which is just what a big blower needs. Just remember that nitrous oxide releases

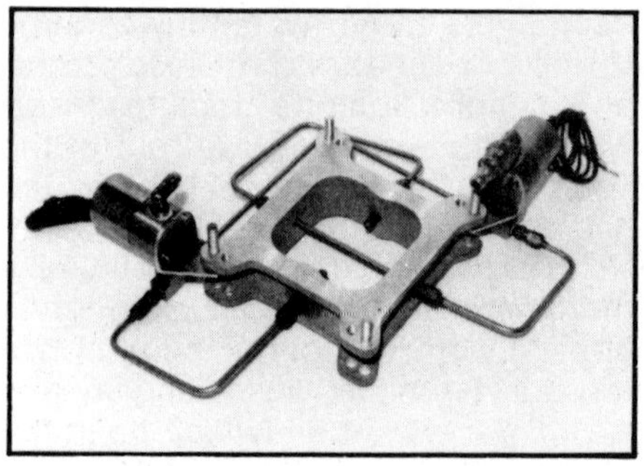

The simplest way to add nitrous oxide injection to any draw-through supercharger application is to install a carb-plate injector unit between the carburetor(s) and the blower. If you do, however, be sure to inform the N_2O manufacturer that you intend to install the system on a blown engine, and have them jet the N_2O and fuel enrichment nozzles accordingly.

extra oxygen in the combustion chamber as well, which requires a certain percentage of increased fuel for proper combustion. A blower nitrous system could be easily set up with slightly larger gasoline nozzles, however, to supply further fuel enrichment at full throttle.

There are a few drawbacks to combining nitrous oxide injection with a supercharger, however. The first, and most obvious, is the added expense. In most situations, by far, the supercharger should be able to provide as much extra horsepower as you would want on the street without need for nitrous. If anything, a considerably less expensive (and simpler to fill) water injector will take care of detonation problems. However, a certain percentage of today's blower market seems to be undaunted by price. If you want to build the "ultimate" blower system, coupling it with nitrous oxide injection could be considered.

Nonetheless, we are speaking more or less theoretically. In practice, the addition of a nitrous oxide injection system to a relatively high-boost blower system could be nothing more than double trouble. It has taken a few years for the nitrous kit manufacturers to get their systems dialed in to the point where they are reliable and effective on *unblown* engines. Recently we saw a couple of years of serious nitrous oxide use on blown fuel and alcohol drag machines, with good results. But not much experimentation has been done yet with nitrous oxide on street blown motors. If the system is not dialed in exactly right, the results could be disastrous in a big hurry. Probably the best approach would be to increase the nitrous oxide dose in small increments, and to initially set up the system to deliver a richer than normal extra fuel shot, just to be on the safe side.

If you are running single or dual four-barrels on top of a Roots-type blower, you could obviously sandwich under-carb nitrous injector plates between the carbs and the blower. Most of the nitrous injector manufacturers could also fabricate a special nozzle plate to mount atop the blower case, or they could even insert the N_2O and extra fuel nozzles directly into the manifold port runners below the blower, where they might be even more effective at cooling the intake charge. However, just remember that the combination of high blower boost, carburetor fuel, nitrous oxide, and extra fuel all mixed up in a big "plenum" like a 6-71 GMC is a highly volatile brew. A backfire under boost (such as from a valve which fails to close) could produce explosive results, to say the least. Only slightly less dramatic would be the effect of a clogged or inoperative fuel enrichment nozzle in the nitrous system, or an otherwise overly lean situation in the engine. In any case, use good quality nitrous equipment, make sure both the nitrous and gasoline lines are filtered, and discuss your plans thoroughly with the N_2O system manufacturer before bolting it onto a supercharged engine.

If you look closely you will spot a nitrous system added to this Paxton-blown Mustang. It consists of the nitrous solenoid (1), the extra fuel solenoid (2), and the fuel and N_2O nozzles (3) which in this case are plumbed into the air pressure bonnet atop the carburetor. Adding a nitrous system to a blow-through set-up like this can be touchy since both the N_2O and enrichment fuel pressure must be balanced against the blower pressure in the bonnet.

AIR CLEANERS

Although you may see countless numbers of supercharged street machines and rods running around with open velocity stacks or injector-type scoops atop the carburetors on their blowers, *you should never operate a street-driven, supercharged vehicle without air cleaners.* It is just plain foolish to do so. In the first place, the only thing that will *wear out* the rotors in a Roots blower, or the impeller in a centrifugal-type blower, is dirty air. Secondly, nothing can ruin a good blower faster than some foreign object ingested through the intake. Remember, that big windmill on top of your motor is just like a vacuum cleaner. It will suck anything near it into the air scoop; and ten thousandths of an inch clearance inside

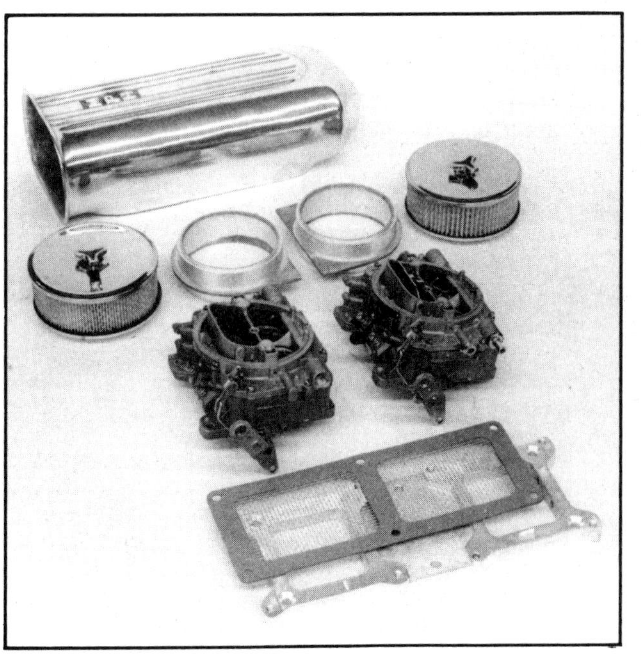

Blower scoops are neat, but they don't filter the air being sucked into the blower and motor. For street set-ups (such as the one from Supercharger Engineering shown here), most manufacturers offer small air cleaners which will fit inside the scoop. However, many of these can be very restrictive or air flow—which defeats the blower's purpose. Racers usually install the wire screen gasket, shown here atop the carb adapter, at the blower inlet, but don't be fooled into thinking this is an air cleaner...it only keeps out the bigger nuts and bolts.

If you must use small air cleaners on your blower carbs (or even if you have room for a larger one), be sure to select a good-flowing element such as the "re-usable" cotton and wire filter shown on the right, such as the ones made by K&N. Such filters are sprayed with a light oil mist, and can be washed and resprayed when they become dirty. A paper filter element, such as the one shown at left, is the next best, but different brands can vary considerably in airflow capacity.

the blower doesn't leave room for rocks, leaves, bird feathers, bugs, or loose nuts and bolts. Although you certainly don't want dirt inside *any* engine, small objects could actually slip through the carburetor, manifold, valves, and combustion chamber of a normally-aspirated engine without causing severe damage. With most blowers—especially Roots-types—even extremely small dirt particles or foreign objects can cause damage to the rotors or cause the blower to seize. Compared to the rest of the blower system, air cleaners are dirt cheap—and they keep out a lot more than just dirt. An added benefit of air cleaners, for those of you who prefer a quiet blower system rather than a screaming banshee, is that they effectively muffle the considerable carburetor "roar" that blowers produce under boost.

That's easy to understand, and adding air cleaner(s) to the carburetor(s) atop you blower is easy enough to do. But there is one problem...one that most hot rodders know, but many blower installers forget. All air cleaners cause some degree of intake air restriction. You have probably read magazine articles about, or actually observed the results of, installing a low-restriction air cleaner element on a normally-aspirated engine. Just taking the top off of a factory air cleaner housing, or turning it upside down, can often cut a full second off quarter-mile times on an otherwise stock late model car.

A normally-aspirated engine is more sensitive to restrictions in the intake system than a blown engine is (since the blower will pump the air through the system). But it makes very little sense to install a blower unit on an engine to increase breathing efficiency, and then cap it off with one or two small, restrictive air cleaners. Not only will such a situation starve the engine for air, but it will also greatly decrease the volumetric and adiabatic efficiencies of the blower, and lead to fuel mixture problems which could ultimately cause damage to the engine.

We seem to have a dilemma here. A blower needs an air cleaner, but an air cleaner combats the purpose of the blower. The solution is to use as large and as efficient an air cleaner as possible on the motor. If you are running a pair of four-barrel carbs atop a 6-71 Jimmy, this may be easier said than done—but the point is that you *don't* want a one-inch-tall, six-inch-diameter paper element filter on each carburetor. If you don't have room for anything else, at least stack a couple of such filter elements on each carb.

Obviously an air filter's ability to flow air is proportional to its size. However, different types of filter elements—and of air cleaner housings—are less restrictive than others. In general, foam-type elements are the most restrictive, paper elements can range in efficiency considerably given brand and quality, and oil-sprayed, cotton and wire mesh elements (such as made by K & N Filters) show superior air flow ratings. For a full discussion of air cleaners and housings, with flow bench test results, see PERFORMANCE WITH ECONOMY, pp 21-25, by David Vizard (available from S-A Design, see page 128).

Centrifugal superchargers—specifically the Paxton—are *very* sensitive to any obstructions or restrictions to the air flow on the blower's inlet side. After years of testing and analysis, Paxton has found a large-diameter, industrial-type air cleaner to be the best for their blower unit. It may not be the prettiest or raciest looking piece, but it is the only one that works effectively with this type of blower. Substituting a smaller, after-market type of air cleaner for the Paxton will not only lower the blower boost, but could even cause lubricating oil to be sucked through the blower shaft seals, causing eventual damage to the blower.

Finally, it should be obvious that the air cleaner(s) should be placed in a fresh air stream, or have outside air directed to them. If your blower sticks through the hood, your carbs will get plenty of fresh air. But under-hood systems, such as a centrifugal blower or a low-profile unit like a Magna Charger, will benefit greatly by being fed cool air rather than drawing from the heated engine compartment. The purpose of a blower is to condense the intake air charge, so you should start with the densest air available; and since heating the air charge is the nemesis of any blower—both in terms of blower efficiency and the detonation level of the engine—lowering the temperature of the air on the intake side of the blower is one sure way to lower the temperature on the outlet side. Underhood temperatures can be very high in any vehicle, but they are often even higher if the engine is supercharged.

GAUGES

You would never start a fresh high performance engine without oil pressure and a water temperature gauges attached. Without gauges, the first indication that something isn't working right is usually a major catastrophe...and then it's too late to fix. The same thing applies to firing up a supercharged engine for the first time without a boost gauge. *Never* install a blower on an engine without installing a good boost gauge in the manifold at the same time. Without one, you really have no idea exactly what the blower is

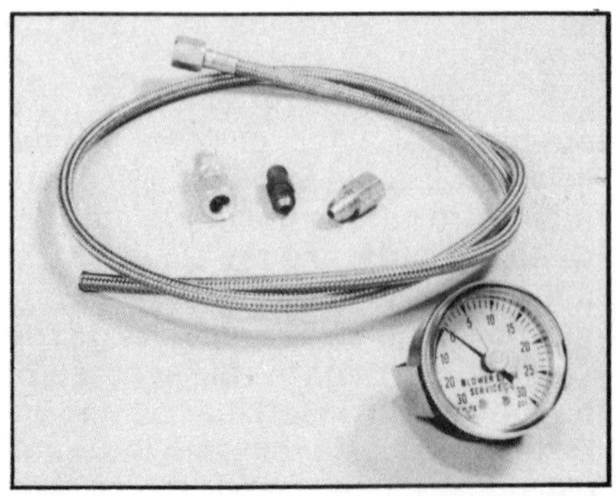

Most blower shops carry a full line of gauges. Blower Drive Service manufactures this "tattle tale" vacuum/boost gauge themselves. The black needle reads manifold vacuum or pressure, while the smaller needle stays set at the highest boost attained during a run.

doing.

Good vacuum/boost blower gauges are available from all of the usual instrument manufacturers, and most shops that sell blowers also carry such gauges. Some even make their own. Just about any air pressure gauge will work, as long as it is calibrated in a range from approximately 0 to 20 or 25 psi, but you might as well buy a gauge made specifically for a supercharged or turbocharged automotive engine, which will indicate both vacuum and pressure on the same dial. Such gauges usually read in inches of mercury on the vacuum side and pounds per square inch (above atmospheric) on the boost side. You can even buy very sensitive, aircraft-type, glycerin-filled (to keep the needle from vibrating) boost gauges, but such a guage is not really necessary on a street machine.

This manifold for an MC-220 Magna Charger has been drilled and tapped for a fitting and hose for a vacuum/boost gauge on the right, and for a temperature gauge on the left. Always insert gauge pickups in a plenum area of the manifold, rather than in one runner, and as far away from manifold walls as possible.

The time to install a boost gauge is *before* you install the blower. The gauge hooks up, via a small hose or air line, to a fitting which must be inserted into the intake manifold (below the blower). Hardly any blower manifolds come drilled for a boost gauge pickup, so you will have to drill and tap the manifold for it before you bolt the manifold onto the engine (you don't want metal chips inside the manifold to be ingested by the motor). Remember to use a pipe thread tap for the hole, and use a bit of teflon tape or other sealant on the threads to make sure no air leaks past the fitting. Where you place the boost gauge pickup doesn't make a whole lot of difference, but it would be best to place it in the plenum wall, rather than in one of the individual port runners.

In the chapter on blower theory we discussed how to compute the density ratio of your blower—or the amount the blower is actually *condensing* the intake charge, rather than just how much it is *compressing* it (the boost gauge just tells you how much the blower is compressing the air). To figure air density in the manifold, you also need an air temperature gauge, the probe for which can be inserted into the manifold. Such a gauge is difficult to find. On a blower dyno test, we attempted to use a couple of different oil temperature gauges for this purpose. One was a Stewart-Warner mechanical oil temperature gauge with a dial reading from 140-325° F; the other was a VDO electric transmission oil temp gauge reading from 90-400° F. Such gauges should work satisfactorily for reading wet air temperature (as found in a manifold). However, at six pounds of boost we were getting no reading on the scale of the S-W gauge, and the VDO gauge seemed to be picking up too much heat transfer from the aluminum manifold to the brass temperature probe. In either case, these gauges responded too slowly to give reliably accurate readings of manifold air temperature at a given boost at a specific rpm on the dyno, since the engine was held at that load only for a short time. The situation might be similar in street driving.

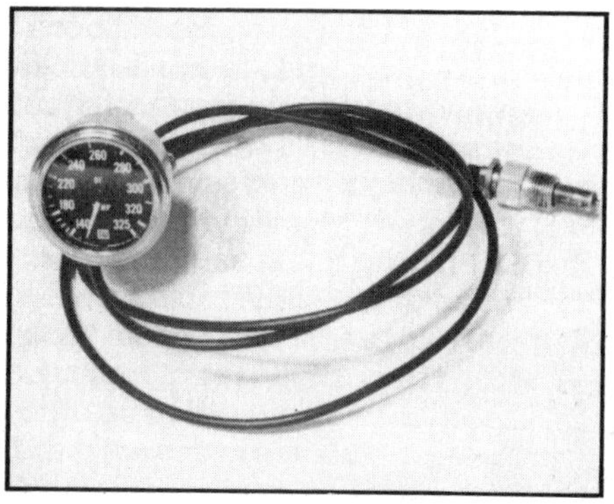

You can use a fluid temperature gauge, such as this S-W transmission temp unit which reads from 125 to 325 degrees, as an economical manifold temperature indicator. However, a thermocouple reads much more quickly.

At higher boost levels, gauges such as these might give better indication of manifold temperature increase with pressure increase. Such gauges—which are readily available and inexpensive— would certainly give you a good ballpark indication of how efficiently your blower is working, even if they are not quite accurate enough to work out density equations. For more accurate, and instantaneous, measurement of intake manifold temperature you will need a good thermocouple (pyrometer) with a scale reading from approximately 60° F to 300° F. A wide range of such gauges and thermocouple probes is available from Omega Engineering of Stamford, Connecticut, among other sources (but be forewarned, they are far from cheap!). You would want an open bead thermocouple which will extend about 2/3 of the way across the manifold opening beneath the blower. These thermocouples do not have any sort of screw-in, airtight fitting, so they would have to be inserted into the manifold through a small brass compression (Farrel) fitting.

A good manifold air temperature gauge will give you an excellent indication of how efficiently your blower is compressing air. By reading both the manifold pressure and the manifold air temperature at a given boost level, you can work out the air density ratio equation given in chapter four by estimating ambient (outside) air temperature, and figuring ambient air pressure to be approximately 14 to 14.7 psi (absolute). However, if you want to calculate the true intake air temperature at the carburetor, you can use an excellent "outside air temperature" gauge made by VDO which reads on a scale from -10° F to 100° F. This gauge uses a 1/8-inch diameter brass sensor which can be threaded directly into the side of an air cleaner or air horn above the carburetor, with a single wire lead to connect to the gauge. And if you want to calculate ambient air pressure more accurately, a variety of barometers are available for mounting in a car.